"*Thomas Berry and the New Cosmology* scrutinizes the position that the relationship between human beings and the earth is in crisis."

*Publishers Weekly*

"[Thomas Berry] has opened our eyes to the reality and the responsibilities of that new and perilous historical situation in which we now live: the ecological age. The recent publication of two of Thomas Berry's essays, together with a chorus of rejoinders from a spectrum of viewpoints, is therefore a very special occasion....

"One can only hope that this thoughtful and thought-provoking collection will prove a vehicle for conveying to a larger audience Thomas Berry's enormously creative and transforming vision."

Donald P. Gray
Manhattan College

"This slim volume is a first formal step in providing a critique for [Thomas] Berry, whose provocative ideas are...worthy of serious consideration."

Lawrence S. Cunningham
*Commonweal*

"The most provocative figure among [the] new breed of eco-theologians is Father Thomas Berry...whose essays have aroused environmentalists like a voice crying for the wilderness."

Kenneth L. Woodward
*Newsweek*

"Members of the Teilhard Association know of [Berry's] probing studies of the human venture and his insistence on the need to go back into the fundamental elements of creation itself. Through the discussions in this book Berry's principles and proposals will be projected into the mainstream of theology today."

*Teilhard Perspective*

"The brevity of this book belies its significance. There has been a hiatus in this line of thought for the ordinary reader since the popular consumption of Teilhard de Chardin in the 1960s. The essential thought of Thomas Berry continues the work of Teilhard. Whereas Teilhard was trying to reconcile Christian belief with scientific data, Berry is trying to reevaluate the resources available to insure survival on the planet."

Emmanuel Sullivan, S.A.
Graymoor Ecumenical Institute
*Journal of Ecumenical Studies*

"Thomas Berry is making very necessary waves in a Catholic church still largely medievally oriented."

Richard Leviton
*Small Press*

"A book to be recommended as a good introduction to Berry's thought and a creation-centered theology."

Pat Feldsien
*Creation*

# Thomas Berry and the New Cosmology

*edited by*
Anne Lonergan
Caroline Richards



XXIII
TWENTY-THIRD PUBLICATIONS
Mystic, Connecticut

**Third printing 1990**

Thomas Berry's essay, "Economics: Its Effect on the Life Systems of the World," was previously published in *Cross Currents* (Winter 1985–1986, Vol. 35, No. 4) under the title "Wonderworld as Wasteworld."

Twenty-Third Publications
185 Willow Street
P.O. Box 180
Mystic, CT 06355
(203) 536-2611



ISBN 0-89622-337-x
Library of Congress Catalog Card Number 87-40528

# CONTENTS

# Thomas Berry
# and the New Cosmology

# *INTRODUCTION:*

# *The Challenge of Thomas Berry*

Anne Lonergan

*Anne Lonergan is a Cenacle sister who has worked in retreat houses and been involved with adult education for many years. She has an M.A. in English from the University of Toronto, and an M.Sc. in Communications from Boston University. She has been studying the work of Thomas Berry for ten years, and has been instrumental in bringing him to Ontario, and developing the library and programs at Holy Cross Centre for ecology and spirituality. As co-editor of this volume, she has contributed the introduction, discussion/ reflection questions and the annotated bibliography.*

This book is a report-in-progress on one of the currents breaking in Christian theology. The question of the relationship between human beings and the earth is fast becoming our most serious, life-threatening problem. Prior to 1980, very little of a serious theological nature was written on this issue. Joseph Sittler and Douglas John Hall were lonely voices in the wilderness. In 1981, Gibson Winter published *Liberating Creation: Foundations of Religious Social Ethics* (New York: Crossroad) and a year later, James Gustafson wrote the first volume of his *Ethics from a Theocentric Perspective* (Chicago: University of Chicago Press). In 1985, Jürgen Moltmann's, *God in Creation: A New Theology of Creation and the Spirit of God,* (San Francisco: Harper & Row), was published.

There he states:

> What we call the environmental crisis is not merely a crisis in the natural environment of human beings. It is nothing less than a crisis in human beings themselves. It is a crisis of life on this planet, a crisis so comprehensive and so irreversible that it can not unjustly be described as apocalyptic. It is not a temporary crisis. As far as we can judge, it is the beginning of a life and death struggle for creation on this earth (p. xi).

A similar perspective has been offered in Hans Jonas' work in philosophical ethics, particularly *The Imperative of Responsibility: In Search of an Ethics for the Technological Age* (Chicago: University of Chicago Press, 1984). By and large, though, the main currents of theology have not addressed the issue with anything like the urgency of Moltmann and Jonas. Periodic reflections on stewardship and a more ecologically responsible Christianity have appeared, and though feminist theologians usually do include ecology as a part of their basic perspective, they have not done it in a systematic way. The urgency of our present crisis is certainly not the basis for reflection in most theological journals.

In the Roman Catholic community today, however, there is a cultural historian who addresses the relationship between human beings and the earth. His name is Thomas Berry. Through articles, lectures, courses, and his own series, called the "Riverdale Papers," Berry has drawn attention to the serious ecological problems of our times and the lack of theological reflection on human/earth relationships.

### *WHO IS THOMAS BERRY?*

Thomas Berry was born in 1914 in North Carolina. He was ordained a Passionist priest in 1942, studied history at the Catholic University of America, and received his doctorate in 1949. From 1948 to 1949 he studied Chinese in the language school at Peking, China, and from 1956 to 1961 he taught at the Institute for Asian Studies at Seton Hall University. From 1960 to 1966 he taught at the Center for Asian Studies at St. John's University in New York.

From 1966 on, he was professor of the History of Religions and director of the History and Religion program in the Theology Department of Fordham University. He has also taught at Columbia University, Drew University, and the University of San Diego.

In 1970, Berry founded the Center for Religious Research in Riverdale, New York. He describes the Center as a place for studying the dynamics of the planet earth, and the role fulfilled by human beings within the total dynamics of the universe. This Center is now the main focus of his work. In 1975, Berry was elected president of the American Teilhard Association, a position he held until 1987

Berry described his intellectual journey in this way:

> I started off as a student of cultural history. I am primarily an historian. What I have to say are the probings of an historian into human affairs in a somewhat comprehensive context. During my university studies I sought to understand the unity and differentiation of human cultures and the dynamism that shaped their sense of reality and value. I studied especially the Chinese language sufficiently to read some of the basic classics. In 1949, I went to China for a year, and when I came back I took up Sanskrit in addition to continuing Chinese studies, so that I could get into the scriptures of Hindu India.
>
> I had begun studying the American Indian world in the 1940's, particularly the Plains Indians. I wished to get beyond the classical civilizations, back into the earlier Shamanic period of the human community. The more I gave to the study of the human venture, the more clearly I saw the need to go back into the dynamics of life itself. I was progressively led back to what I call the study of the earth community, including its geological and biological as well as its human components. I call myself a geologian.

This book includes Berry's "Twelve Principles for Understanding the Universe" and two of his papers: "The Earth: A New Context for Religious Unity," and "Economics: Its Effect on the Life Systems of the World." Both papers provide in-depth insights into Thomas Berry's ideas and the challenge he presents. Though these papers may not represent the full scope of Thomas Berry's remarkable range of interests, they are, according to contributor Donald Senior, "good examples of his current concerns and theological style."

### *THE CHALLENGE PICKED UP*

Berry's challenge has been picked up by a number of his contemporaries. These people, primarily theologians, have read the "Twelve Principles" and the two papers included in this book, and have reacted to them. They raise many fundamental issues that others, particularly in theological circles, might also want to raise with Berry.

Contributors include: Donald Senior from Catholic Theological Union, Chicago; Margaret Brennan from Regis College, Toronto; Gregory Baum, from McGill University, Montreal; James Farris of Knox College, Toronto; Stephen Dunn of St. Michael's College, Toronto; Brian Swimme, program director at the Institute in Culture and Creation Spirituality in Oakland, California; and Caroline Richards of Earlham College, Indiana.

The questions following each paper are meant to stimulate discussion and to help readers probe the issues involved more personally. We would suggest that you concentrate on those questions most relevant to your concerns, and if time permits go on to others that interest you. You will note that questions are sometimes drawn directly from the text, but at other times they reflect the concerns of participants in adult education programs with Thomas Berry.

Anne Lonergan and Caroline Richards have edited these papers and provided introductions to the material. Our special thanks to all the contributors for their gracious cooperation, and friendly, yet frank contributions. We especially thank them for allowing us to take them "off their usual space" into areas that are not yet in the mainstream of theology. Their responses indicate the measure of profound commitment each has to the task of authentic theology in the present and for the future.

# 1

# *ECONOMICS:*

## *Its Effect on the Life Systems of the World*

Thomas Berry

Economics as a religious issue can be dealt with in different ways. One way is to begin with the religious quest for justice. In this context we have a special concern that the well-being of the society be shared by all, that the basic life necessities be available to the less privileged. Such an approach emphasizes our social and political responsibilities to see that the weak and less gifted are not exploited by the strong and the competent.

This moral-religious critique generally concerns itself with the issue of a capitalist market economy that neglects its social responsibilities. The remedy offered, in accord with biblical and moral principles, is to incorporate everyone into the functioning and benefits of the economy. Admirable as this approach may be, it brings about only temporary improvement since the more basic difficulty may not be the social issue but the industrial economy itself, which is not a sustainable economy.

Another way of dealing with economics as a religious issue is to begin with the present economy and inquire into its deeper implications from within its own functioning. This is the manner of procedure we will be using here. We will begin with a few observations concerning the reality of the present economy and its capacity to sustain itself. We will also look at its consequences for the well-being of the human community and for the life systems of the earth upon which a sustainable economy depends in a very direct manner.

The reality of our present economy is that it does not bode well for either the human community or even for the planet itself in its most basic life systems. Economic dysfunction is generally expressed in terms of deficit expenditure. Income does not balance outflow. In the natural world there exists an amazing richness of life expression in the ever-renewing cycle of the seasons. There is a minimum of entropy. The inflow of energy and the outflow are such that the process is sustainable over an indefinite period of time. So long as the human process is integral to these processes of nature, so long is the human economy sustainable into the future.

The difficulty comes when the industrial mode of our economy disrupts the natural processes, when human technologies are destructive of earth technologies. In such a situation the productivity of the natural world and its life systems is diminished. When nature goes into deficit, then we go into deficit. When this occurs to a limited extent on a regional basis it can often enough be remedied. The difficulty is when the entire planetary system is affected. The earth system is most threatened when the human economy goes out of balance and frantic efforts toward a remedy lead to a reckless plundering of the land, spending our capital as our interest diminishes.

If we look at the specific data available in the United States economy we find that there is now a GNP of over 3 trillion dollars. There is a national debt of 1,800 billion dollars, an annual budgetary deficit of some 200 billion, an infrastructure disintegration requiring repairs of 750 billion, an annual trade deficit of over 100 billion, 200 billion in Third World financial loans unlikely to be repaid, and an annual military expenditure of 300 billion. All of these can be considered financial deficits.

### *WHAT OF THE EARTH DEFICIT?*

But seldom does anyone speak of the earth deficit, the deficit involved in the closing down of the basic life system of the planet through abuse of the air, the soil, the water, and the vegetation. As we have indicated, the earth deficit is the real deficit, the ultimate deficit, the deficit with some consequences so absolute as to be beyond adjustment from any source in heaven or earth. Since the earth system is the ultimate guarantee of all deficits, a failure here is a failure of last resort. Neither economic viability nor improvement in life conditions for the poor can be realized in such circumstances. These can only worsen, especially when we consider the rising population levels throughout the developing world.

This deficit in its extreme expression is not only a resource deficit but the death of a living process, not simply the death of *a* living process but of *the* living process (a living process that exists, so far as we know, only on the planet earth). This is what makes our problems definitively different from those of any other generation of whatever ethnic, cultural, political, or religious tradition, or of any other historical period. For the first time we are determining the destinies of the earth in a comprehensive and irreversible manner. The immediate danger is not *possible* nuclear war but *actual* industrial plundering.

Economics on this scale is not simply economics of the human community; it is economics of the earth community in its comprehensive dimensions. Nor is this a question of profit or loss in terms of personal or community well-being in a functioning earth system. Economics has invaded the earth system itself. Our industrial economy is closing down the planet in the most basic modes of its functioning. The air, the water, the soil are already in a degraded condition. Forests are dying on this very continent. The seas are endangered. Aquatic life-forms in lakes and streams and in the seas are contaminated. The rain is acid rain.

So the litany goes on. The United States loses over four billion tons of topsoil each year. The great aquifers of the Plains region are diminished beyond their capacity for refilling. Our industrial agriculture is no longer participating in the productive cycles of the natural world; it is the extinction of the very conditions on which these productive cycles depend.

While it is unlikely that we could ever extinguish life in any absolute manner, we are eliminating species at a rate never before known in historic time and in a manner never known in biological time. Destruction of the tropical rain forests of the planet will involve destroying the habitat of perhaps half the living species of earth. Although its strictly economic implications have still not been worked out, it should be clear: An exhausted planet is an exhausted economy.

### *WELL-DOCUMENTED STUDIES*

The earth deficit in its resources and in its functioning has been documented in a long series of specialized studies and in more general evaluations in ever-increasing volume over the past 20 years. The first thorough scientific study of the situation was that of Rachel Carson who described the chemical poisoning of the land and the killing of its life systems in her 1962 book, *Silent Spring* (Boston: Houghton Mifflin). In 1970, Paul Ehrlich edited a comprehensive study entitled *Ecoscience: Population, Resources, Environment* (San Francisco: W.H. Freeman). Then in 1972 came the comprehensive survey of the planet earth as a complex of life systems, *The Limits to Growth* (New York: Universe Books), edited by Donella Meadows with several others, a work based on the earlier *World Dynamics* by W. Forrester (Cambridge, Mass.: MIT Press, 2nd ed. 1972). In 1976, an unappreciated work originally published in 1952 was republished, a work by Edward Hyams entitled *Soil and Civilization* (New York: Harper & Row). This is an extraordinary study of the difficulties encountered in establishing sustainable human relations with the land in various civilizations, even suggesting that the destruction of the natural environment in the Mediterranean world by the classical civilizations contributed significantly to their decline.

In 1980 came a second comprehensive survey of the planet earth entitled, *Global 2000: A Report to the President,* edited under the direction of Gerald Barney (Washington, D.C.: Council on Environmental Quality and the Department of State). Then in 1981 a valuable survey of this report and four other global reports was given by Magda Cordell McHale in her book, *Ominous Trends and*

*Valid Hopes* (Minneapolis: Hubert Humphrey Institute of Public Affairs, 1981).

One of the most helpful of these general studies was published by Norman Myers in 1984, *Gaia: An Atlas of Planet Management* (New York: Doubleday). In that same year, Lester Brown, with the resources of Worldwatch Institute, started an annual publication entitled *State of the World.* So far two issues are available, 1984 and 1985. The listing of specialized studies would be endless. There are depth inquiries into why this assault on the earth is taking place, such as Carolyn Merchant's study, *The Death of Nature* (New York: Harper & Row, 1980). There are special studies in different fields such as the studies of agriculture by Rodale Press. All of these indicate that the planet cannot long endure present modes of human exploitation.

Until recently both textbook economics and corporation practice have ignored the implications of such data or have given it minimal attention. Such deficits were simply external or unreal costs of doing business, costs that were not entered into the bookkeeping records until social protest brought about environmental impact statements, limits on pollution of the environment, clean-up of waste sites, and liability for personal damage resulting from toxic disruption of the basic life systems. Even the existence of such clean-up needs should be telling us something: The industrial system itself in its present form is a failing system. Yet we can be sure that whatever fictions exist in Wall Street bookkeeping, the earth is a faithful scribe, a faultless calculator, a superb bookkeeper; we will be held responsible for every bit of our economic folly.

Only now do we begin to consider that there is an economics of the human as a species as well as an economics of the earth as a functional community. We have just begun to realize that the primary objective of economic science, of the engineering profession, of technological invention, of industrial processing, of financial investment, and of corporation management, must be the integration of human well-being within the context of the well-being of the natural world. This is the primary purpose of economics. Only within the ever-renewing processes of the natural world is there any future for the human community. Not to recognize this is to make economics a deadly affair.

### *THE MYTH OF ECONOMICS*

The exploitation of the earth was and still is experienced by economists not as deterioration of the planet or as a new mode of exhaustion of the planet but as an extension of the emergent creative process leading to a kind of wonderworld existence. This is "progress," a belief so entrancing for the modern world that doubt about its validity is not permitted. Even though this belief has long since been severely critiqued and its limitation indicated, it remains the functional basis of our economy. The GNP must increase each year. Everything must be done on a larger scale, with little awareness of the inbuilt catastrophy involved in the exponential rate of increase. However rational modern economics might be, the *dynamic* of economics is visionary. This dynamic is a visionary commitment supported by myth and a sense of having the magic powers of science to overcome any difficulty encountered when human processes reach their limits.

This visionary approach can be seen in the new surge of the industrial economy, the rising level of stock market quotations, the shifting of currency values, the formation of the great conglomerates, the giant corporation mergers, the new mystique of the entrepreneur. This last item is described in the recent best-seller *In Search of Excellence* (New York: Harper & Row, 1982) by Thomas Peters and Robert Waterman, Jr., and given its archetypal model in the autobiography of Lee Iacocca.

Herman Kahn and Julian Simon have argued in defense of this myth of process with severe criticism of forebodings concerning the national or world economy such as those presented here. Herman Kahn resented especially the rejection of the idea of limitless progress and the dangers of exponential rates of increase presented in *Limits to Growth* and *Global 2000: A Report to the President.* He encouraged us to continue our established way into the future, confident that our scientific insight, technological competence, and economic discernment would lead us on into an even better life situation. The best summaries of this position of ever-continuing process can be found in the book by Julian Simon entitled, *The Ultimate Resource* (Princeton: Princeton University Press, 1982) and the one he edited, *The Resourceful Earth* (Oxford: Basil Blackwell, 1984). These works argue that every generation in modern times has lived better than the prior generation, that there is no serious problem, we must not lose our nerve, science can resolve our difficulties. Presently there is a glut of food in America, an increase in land brought under cultivation throughout the world, so why worry about the loss of topsoil?

Thus the mythic drive continues to control our world even though so much is known about the earth: its limited resources, the interdependence of life systems, the delicate balance of its ecosystems, the consequences of disturbing the atmospheric conditions, of contaminating the air, the soil, the waterways and the seas, the limited quantity of fossil fuels in the earth, the inherent danger of chemicals discharged into natural surroundings.

Although all this has been known for generations, neither the study nor the commercial-industrial practice of economics has shown any capacity to break free from the mythic commitment to progress, or any awareness that we are in reality creating wasteworld rather than wonderworld. This mythic commitment to continuing economic growth is such that none of our major newspapers or magazines considers having an ecological section in each issue—equivalent to the sports section, or the financial section, or the arts section, or the comic section, or the entertainment section—although the ecological issues are more important than any of these, even more important than the daily national and international political news. The real history that is being made is inter-species and human-earth history, not inter-nation history. Our real threat is from the retaliatory powers of the abused earth, not from other nations.

If this assault on the earth were done by evil persons with destructive intentions it would be understandable. The tragedy is that our economy is being run by persons with good intentions under the illusion that they are only bringing great benefits to the world and even fulfilling a sacred task on the part of the human community. "We bring good things to life." "Progress is our most important product." "Fly the friendly skies." These are millennial dreams for moving on into new frontiers of economic accomplishment for the fulfillment of the high purposes of the universe itself.

Nor has the real situation been appreciated by social reformers or by those concerned with the needs of the poor and dispossessed. These, whether socialist or capitalist in orientation, wish mainly to enable the poor to find their place in the industrial world. Whether this is privately or socially controlled, the industrial process itself is generally accepted.

Nor have our moral theologians been able to deal with our abuse of the natural world. After dealing with suicide, homicide, and genocide, our western Christian moral code collapses completely: It cannot deal with biocide or geocide. Nor have church authorities made any sustained protest against the violence being done to the planet.

### *A SLOWLY EMERGING SENSE*

The new sense of what economics is all about has emerged from the naturalist Aldo Leopold in his essay, "A Land Ethic," from an independent biologist, Rachel Carson, in *Silent Spring,* and from the economist, Nicholas Georgescu-Roegen, in *The Entropy Law and the Economic Process* (Cambridge: Harvard University Press, 1971). Georgescu-Roegen, in particular, had a profound sense of the economic implications of the second law of thermodynamics. Before his time no modern economic system yet had any realization of the earth system itself as the primary functional context of life in all its aspects. Every modern economic system from the mercantile and physiocrat theories of the 17th and 18th centuries, to the supply-demand theories of Keynes is homocentric and exploitive. The natural world is considered as a resource for human utility, not as a functioning community of mutually supporting life systems within which the human must discover its proper role.

The basic critique of Georgescu-Roegen is that economists were caught in a mechanistic world that could be understood simply from within its own economic data. So with this model, derived from Newtonian cosmology, the economists in their theories and the corporations with their practice sought to manage the economic world from within such a limited context. Economics was a closed process of commercial transactions with reference only to the production and exchange of goods. As Georgescu-Roegen indicates, "Economists do speak occasionally of natural resources. Yet the fact remains that, search as one may, in none of the economic models in existence is there a variable standing for nature's perennial contribution" (*The Entropy Law and the Economic Process,* p.2). He also notes, "The fact that biological and economic factors may overlap and interact in some surprising ways, though well established, is little known among economists" (p. 317).

Even now corporations feel imposed upon when they are required to make environmental impact statements concerning their intrusion into the natural world, when they are required to refrain from scattering industrial waste over the land, to indicate to their workers the toxic nature of the materials they are working with, or when they are required to list the chemical contents of their products.

There is a certain pathos in social reform movements and in the efforts made to improve the living conditions of the impoverished within the context of such a dysfunctional and non-sustainable economy. This is understandable however since life necessities, air and water, food, clothing and shelter, are demanded presently. Tomorrow is too late. Whatever the existing economy, human needs must be supplied, even though food today for the few may be starvation tomorrow for the many. This means jobs within the existing context. No immediate alternative seems available.

Even so, an awareness should exist that the present system is too devastating to the natural fruitfulness of the earth to long supply human needs. Alternative programs are being elaborated and becoming functional. If the moral norm of economics is what is happening to the millions of persons in need, then these more functional economic developments are required not only by those excluded from the present system but by the entire nation community, by the entire human community, and by the entire earth community.

## THE DIVERSITY OF CREATURES

This is not socialism on the national scale, nor is it inter-nation socialism. It is planetary socialism. It is a socialism based on the *Summa Theologica* of Thomas Aquinas, (Part I, Question 47, article 1), where he deals with the diversity of creatures. Beyond planetary socialism he proposes an ultimate universe socialism where he says that because the divine goodness

> ...could not be adequately represented by one creature alone, He produced many and diverse creatures, that what was wanting to one in the representation of the divine goodness might be supplied by another. For goodness, which in God is simple and uniform, in creatures is manifold and divided; and hence the whole universe together participates the divine goodness more perfectly, and represents it better than any single creature whatever.

From this we could argue that the community of all the components of the planet earth is primary in the divine intention. Even biologically it is evident that the well-being of the earth itself is a primary consideration if there is to be a well-being of the various components of the earth. The trees of the Appalachian mountains will not be healthy if the rain is acid rain. Nor will the soil be fertile, nor will humans have their proper nourishment. Nor will the human imagination be activated to its grand poetic visions, nor will our sense of the divine be so exalted if the earth is diminished in its glory. It is all quite clear. If we pull the threads, the fabric falls apart, the human fabric in particular, in both its religious and its economic aspects. We come to the essential problem of economics as a religious issue when we consider that the present threat to both economics and religion is from a single source, the disruption of the natural world. If the water is polluted it can neither be drunk nor used for baptism, for it no longer symbolizes life; it is a symbol of death.

Obviously, then, economics and religion are two aspects of a single earth process. If the economy is more immediately the cause for disruption of the natural world, the more ultimate sources for this mode of economic activity may be found in the religious-cultural context within which our present economy emerged.

This may well be the reason why at this time when threatened in the very source of our sense of the divine and in our sacramental forms there is no sustained religious protest or moral judgement concerned with the industrial assault on the earth, the degradation of its life systems, or the threatened extinction of its most elaborate modes of life expression. Even more important: Why did this process develop in a civilization that emerged out of a biblical-Christian matrix?

This most urgent theological issue, so far as I know, has never been dealt with in any effective manner although the accusation has been made by Lynn White, Jr., that Christianity bears "an immense burden of guilt" for the present ecological crisis. A long list of answers have been written, a few by theologians, but mostly brief articles not entirely convincing because of their inadequate consideration of those dark or limited aspects of Christianity that made our western society liable to act so harshly toward the natural world.

### *EXAGGERATED RELIGIOUS ORIENTATION*

Any thorough study of our biblical-Christian traditions in their historical realization reveals immediately a number of religious orientations that have taken possession of western consciousness to an exaggerated degree. The more intensely the religious dedication the greater the imbalance tends to be.

The first of these religious orientations is the biblical commitment to a transcendent personal monotheistic concept of deity with severe prohibition against any worship of divinity resident in nature. By absorbing the original pervasive presence of the divine throughout the natural world and constellating the divine in a strictly transcendent mode, the natural world was to some extent despiritualized and desacralized. The very purpose of Genesis was to withdraw Israel from the Near Eastern orientation. Whatever the benefits of such diminishment of the divine in the natural world, it rendered the world less personal, less subject; it became something seductive, more liable to be treated as object.

Secondly, the redemption experience became the dominant

mode of Christian consciousness to the diminishment of attention to creation experience. A general sensitivity to the natural world and to cosmology remained up through the medieval period. But then during the 14th century, after the Black Death, an overwhelming commitment to redemption controlled the Christian experience. The Apostles' Creed as well as the Nicene Creed both glide lightly over the creation reference. The Council of Trent was so caught up in redemption issues that it had nothing at all to say about creation. Until our times, creation has never been the basic issue. It was simply there; it was beyond discussion. Nature gradually disappeared from Christian consciousness.

A third aspect of the Christian life orientation that brought about our present alienation from the natural world is the Christian emphasis on the spiritual nature of the human over against the physical nature of the other creatures. This attitude was strengthened by the influence of Platonic philosophy from the Hellenic world. Human perfection was thought of in terms of detachment from the phenomenal world in favor of the divine eternal world presented as our true destiny. As this emphasis has developed since the 16th century, we have more and more thought of the natural world as object to be exploited to our rational satisfaction, to our aesthetic enjoyment, or to our utilization as natural resource. In any case the natural world is ultimately considered as object, as not possessing rights or subjectivity or legal status, certainly not constituting with the human a single earth community as a segment of the comprehensive universe community.

A fourth aspect of Christian tradition that made possible such devastation of the natural world as we presently witness it is the expectation of an infra-historical millennial period in which the human condition would be overcome; peace and justice would pervade the land under the spiritual reign of the saints and by the infallible efficacy of divine power. It would be difficult to exaggerate the importance of this belief in western tradition and now in the world tradition, since it has, in a transformed version, been turned into the secular doctrine of unfailing progress in human affairs. This mythic belief has evoked the enormous energies required for creating the industrial world such as we know it.

Because the human condition was not overcome by spiritual power or divine intervention, humans have since the time of Francis Bacon been determined to bring this better world into being through the scientific, technological, industrial, and corporative enterprise. If this requires the total despoiling of the earth to achieve such transformation, then so be it.

While none of these Christian beliefs individually is adequate as an explanation of the alienation we experience in our natural setting, they become convincing in their totality in providing a basis for understanding how so much planetary destruction has been possible in our western tradition. We are radically oriented away from the natural world. It has no rights; it exists for human utility, even if for spiritual utility.

### *THE DIVINE IN NATURE*

Because our sense of the divine is so extensively derived from verbal sources, mostly through the biblical Scriptures, we seldom notice how much we have lost contact with the revelation of the divine in nature. Yet our exalted sense of the divine comes from the grandeur of the universe, especially from the earth in all the splendid modes of its expression. Without such experience, we would be terribly impoverished in our religious and spiritual development, even in our emotional, imaginative, and intellectual development. If we lived on the moon, our sense of the divine would reflect the lunar landscape, our imagination would be as desolate as the moon, our emotions lacking in the sensitivity developed in our experience of the sensuous variety of the luxuriant earth. If a beautiful earth gives us an exalted idea of the divine, an industrially despoiled planet will give us a corresponding idea of God.

But even this sense of the divine tends to draw us away from the sacred dimension of the earth in itself. This is not exactly the divine immanence. We go too quickly from the merely physical order of things to the divine presence in things. While this is important, it is also important that we develop a sense of the reality and nobility of the natural world in itself. Aquinas dedicated his efforts in great part to defending the reality and goodness and efficacy intrinsic to the natural world. The natural world is not simply object, not simply a usable thing, not an inert mode of being awaiting its destiny to be exploited by humans.

The natural world is subject as well as object. The natural world is the maternal source whence we emerge into being as earthlings. The natural world is the life-giving nourishment of our physical, emotional, aesthetic, moral, and religious existence. The natural world is the larger sacred community to which we belong. To be alienated from this community is to become destitute in all that makes us human. To damage this community is to diminish our own existence.

If this sense of the sacred character of the natural world as our primary revelation of the divine is our first need, our second need is to diminish our emphasis on redemption experience in favor of a greater emphasis on creation processes. Creation, however, must now be experienced as the emergence of the universe as a psychic-spiritual as well as a material-physical reality from the beginning. We need to know the great story of the universe in its four phases of emergence: the galactic story, the earth story, the life story, the human story.

## *CONSCIOUS SELF-AWARENESS*

We need to see ourselves as integral with this emergent process, as that being in whom the universe reflects on and celebrates itself in conscious self-awareness. Once we begin to experience ourselves in this manner, we immediately perceive how adverse to our own well-being physically and spiritually as well as economically is any degradation of the planet.

A third need is to provide a way of thinking about "progress" that would include the entire earth community. If there is to be real and sustainable progress it must be a continuing enhancement of life for the entire planetary community. It must be shared by all the living from the plankton in the sea to the birds above the land. It must include the grasses, the trees and the living creatures of the earth. True progress must sustain the purity and life-giving qualities of both the air and the water. The integrity of these life systems must be normative for any progress worthy of the name.

Already these three commitments—to the natural world as revelatory, to the earth community as our primary loyalty in a biocentric rather than a homocentric orientation, and to the progress of the community in its integrity—these three commitments constitute the new religious-spiritual context for carrying out a change of direction in human-earth development. For indeed this is the order of the magnitude of the task that is before us. If the industrial economy—in its full effects which has well-nigh done us in, along with a major part of the entire life community—has been such a massive revolutionary experience for the earth and the entire living community, then the termination of this industrial devastation and the inauguration of a more sustainable lifestyle must be of a proportionate order of magnitude.

The industrial age itself, as we have known it, can be described as a period of technological entrancement, an altered state of consciousness, a mental fixation that alone can explain how we came to ruin our air and water and soil and to severely damage all our basic life systems under the illusion of "progress."

But now that the trance is passing, we have before us the task of structuring a human mode of life within the earth complex of communities. This task is now on the scale of "reinventing the human," since none of the prior cultures or concepts of the human can deal with these issues on the scale required.

### *A NEW LIFE PROGRAM*

Fortunately, a number of creative persons have, over the past twenty years, identified the main features of the new life program. Among the first persons to be mentioned as authentic guides to the future is Edward Schumacher whose little book, *Small Is Beautiful* (New York: Harper & Row, 1975), constitutes a first principle of absolute importance that technologies should be appropriate to the function to be carried out. This simple principle pointed out one of the most obvious mistakes of western engineering enterprises: the lack of proper relation of the technology and the human need to be fulfilled. This exaggerated scale of our present energy systems was later pointed out by Amory Lovins in *Soft Energy Paths* (New York: Harper & Row, 1979). The vast corporative structures seem determined to build ever bigger energy-consuming constructions when much simpler, more efficient, and less expansive methods and materials are available and infinitely more benign to the earth.

This question of human-scale was later developed by Kirkpatrick Sale. So too, The Solar Energy Research Institute developed its program, *A New Prosperity*, based on energy systems less polluting, less costly. Another contribution along these lines is found in *Progress as if Survival Mattered*, by Friends of the Earth (Ellenwood, Georgia: Alternatives, 1981), a book of extensive practical value.

One of the main directions into the future must also be concerned with human habitation. Here much important work has been done by architects in association with others concerned with establishing more intimate, more functional and more viable communities. Gary Coates is among the most outstanding in this general area, especially with the information he has presented of community efforts recently underway in various parts of the country. His study, *Resettling America* (Andover, Mass.: Brick House Pub. Co., 1981), presents a number of such community projects. There we find such titles as *Rural New Towns for America; An Ecological Village; The Rise of New City-States; Urban Agriculture; Goals for Regional Development*. These are all presently in progress.

Out of their own experience with the New Alchemy Institute on Cape Cod and in alliance with others interested in the subject, John and Nancy Todd have presented us with a vision and the drawings for *The Village as Solar Ecology* (East Falmouth, Mass.: New Alchemy Institute, 1980). And, in agriculture there are a multitude of new developments. The experience and writings of Wendell Berry on this subject come from a person who is farmer, writer, thinker, and teacher. His critique of industrial agriculture, the absurdities that have emerged from it, its destructive impact on human life, are among the most effective studies available. But even more important is his presentation of the mystical bond between humans and the earth in his work, *The Unsettling of America: Culture and Agriculture* (San Francisco: Sierra, 1977).

Then there is the work of John and Nancy Todd in their bio-shelter projects; Wes Jackson in his work at The Land Institute and his recovery of genetic strains of grasses of the Plains region; Bill Mollison in his Permaculture program being carried out at the Permaculture Institute of North America on the Northwest coast; Robert Rodale and the Regeneration project being directed from Emmaus, Pennsylvania, The Farmland Foundation. These are only a selection out of hundreds of projects being carried out at the present time.

In alliance with these projects there are the fifty-some ecologically oriented organizations in the United States that joined together a few years ago to defend the North American continent from government supported corporation abuse. These are all efforts outside the professional or official establishments. Objections exist not only in economics and industry, in boardrooms and research laboratories, but also in education, law, medicine, and religion. Having become part of the bureaucratic process, all these find serious difficulty thinking about, much less adapting to, change of the magnitude I am suggesting.

For the past hundred years the great technical engineering schools, the research laboratories, and the massive corporations have dominated the North American continent and even an extensive portion of the earth itself. In alliance with governments, the media, universities, and the general approval of religious groups, they have been the main instruments for producing acid rain, hazardous waste, chemical agriculture, the horrendous loss of topsoil, wetlands, and forests, and a host of other evils the natural world has had to endure from human agency. The corporations should be judged by their own severe norms of exactly what they have produced, the kind of a world they have given us after a century of control.

### *A NEW SURGE OF ACTIVITY*

Feeling threatened now by the rising movements for change, corporations are seeking to strengthen their position. The new surge of economic activity throughout the world, the rising stock market, the enormous assets accumulated that now rise toward the hundred billion level, the global extent of assets, the new ease of information gathering, analysis and communication, all this functions like a great exhilarating wave of achievement in corporate consciousness. A feeling of euphoria pervades the business world, a euphoria shared in by the investing public, although there are deep forebodings concerning the future of the industrial enterprise.

The absolute limitations indicated some years ago in the Club of Rome report and the absurdity of exponential rates of growth begin to assert themselves. Since growth is the central fact of contemporary economics, a sudden confrontation with the inherent limits of earth development is coming as a sobering experience indeed. Before this moment arrives however, an extensive series of confrontations can be expected. Already these are occurring in every aspect of human endeavor, in all our institutions, professions, and activities.

Earlier successes in environmental legislation in the 1970s in setting up regulatory agencies and norms concerning the quality of air and water and waste disposal, were the occasion for later resentment by industry and corporative enterprise. Increasing difficulty is experienced in meeting standards. But even greater difficulty is experienced in enforcing any standards. In many instances the situation is fairly clear concerning what needs to be done. The difficulty for government, for industry, and for the citizenry is accepting the consequences of the changes required, for we are involved in changes in the deep structure of our sense of reality and of value, as well as in the practical adaptation to lifestyles that make less extravagant demands on the environment.

We must be aware of how difficult our present situation is for everyone, even when there is a willingness to deal effectively with the issues before us. The scientific determination of acceptable standards in environmental purity is enormously difficult. The technologies for meeting these standards and their cost to the society, the sensitivity of the citizenry: all these are difficult. By entering into an industrial economy we may have taken on a task beyond human capabilities for both judgement and execution. The arrogance of our engineering intrusion into nature is only now being manifest, as well as our arrogance and our naivete concerning our rational skills and our inventive genius.

The difficulty is that the arrogance continues even when its deleterious consequences are so evident. The Herman Kahns and Julian Simons wish us to press on even further into the wasteworld that we are creating. President Reagan foils the efforts of the Environmental Protection Agency to fulfill its official mandate. These are the attitudes that evoke such initiatives as the Greenpeace Movement on the seas and the Earth First Movement on the land. There are the lawsuits initiated by the National Resources Defense Council and by some of the larger ecologically oriented organizations. Extensive lobbying is going on in state legislatures and in congress. Along with these are the great variety of spontaneous protest movements and a great volume of newletters, reports, and periodicals that have taken a stance and are making demands on industrial and political establishments. These presently are finding a way to sustained influence on the society through Green Movements, and in Green Politics. In some countries Green Parties are beginning to function.

### *A SIGNIFICANT MOVEMENT*

Among the most significant of these general social movements, the one with most efficiency may ultimately be the Bioregional Movement. This movement, especially strong just now in North America, is based on a realization that the earth expresses itself not in some uniform life system throughout the globe but in a variety of regional integrations, in bioregions, which can be described as identifiable geographical areas of interacting life systems that are relatively self-sustaining in the ever-renewing processes of nature. As we diminish our commitment to our present industrial context of life with its non-self-renewing infrastructures, we will need to integrate our human communities with the ever-renewing bioregional communities of the place where we find ourselves.

We need to re-align human dwelling and human divisions of the earth with the biological regions. This will provide a primary biological identity rather than a primary political identity. Our cultural development within this context could have a new vigor derived from such intimate association with the dynamism and artisitic creativity of nature. Much more can be said of the bioregional movement in both its confrontational and in its creative aspects. Its power is in its integration of the human within the cosmological process. In this manner it achieves what a number of new age writers fail to achieve.

These writers wish to be totally realistic and yet hopeful; they even see an exciting future taking shape across the board, as it were, in the events of our times. They see new creative attitudes in the physical, biological, and psychological sciences as well as in economics, politics, social structure, and religion. Such is the presentation of Marilyn Ferguson in *The Aquarian Conspiracy* (Los Angeles: J.P. Tarcher, 1981). We find a similar attitude in *Megatrends* by John Naisbitt (New York: Warner Books, 1983) and *The Third Wave* (New York: Morrow, 1980) by Alvin Toffler. The latter two present amazing amounts of information on every aspect of contemporary life. Peter Drucker with his concentration on the managerial role contributes considerably to our understanding of the controlling processes of our corporative institutions. Robert Heilbrunner gives us a more profound insight into the governing principles and ideals of our economy.

But all fail ultimately in judging the present, and in outlining a program for the future, because none of them are able to present their data consistently within a functional cosmology. Neither humans as a species, nor any of our activities, can be understood in any significant manner except in our role in the functioning of the earth and of the universe itself. We come into existence, have our present meaning, attain our destiny within this numinous context, for the universe in its every phase is numinous in its depths, is revelatory in its functioning, and in it human expression finds its fulfillment in celebratory self-awareness. Neither the psychological, sociological, or theological approaches are adequate. The controlling context must be a functional cosmology.

### *THE ROLE OF RELIGIONS*

At this time the question arises as regards the role of the traditional religions. My own view is that any effective response to these issues requires a religious context but that the existing religious traditions are too distant from our new sense of the universe to be adequate to the task that is before us. We cannot do without the traditional religions, but they cannot presently do what needs to be done.

We need a new type of religious orientation. This must, in my view, emerge from our new story of the universe. This constitutes, it seems, a new revelatory experience which can be understood as soon as we recognize that the evolutionary process is from the beginning a spiritual as well as a physical process. The difficulty so far has been that this story has been told simply as a physical process. Now, however, the scientists themselves are awakening to the wonder and the mystery of the universe, even to its numinous qualities. They begin to experience also the mythic aspect of their own scientific expressions. Every term used in science is laden with greater mythic meaning than rational comprehension. Thus science has overcome its earlier limitations out of its own resources. Brian Swimme, a physicist and cosmologist, and whose paper appears later in this book, notes concerning the scientists: "Their experiences with the most awesome realities of the universe revealed a fantastic dimension that exceeded, exploded, and destroyed the language of the everyday realm." He notes further, "To speak of the diaphanous quality of matter, or to speak of the cosmic dawn of the universe is to treat questions that every culture throughout history has confronted." Just as the shaman in tribal societies and contemplative saints in the classical religious traditions, so the scientist "has returned to the larger culture with stories awesome and frightening, but stories that serve to mediate ultimate reality to the larger culture" (*Teilhard Perspective,* July 1983, Vol. 16, no. 1, p.4).

We are entering into a period that might be identified as the period of Third Mediation. For a long period the divine-human mediation was the dominant context, not only of religion but of the entire span of human activities. Then for some centuries among industrial classes and nation-states a primary concern has been inter-human mediation. In the future, the other two mediations will be heavily dependent on our ability to establish a mutually enhancing human-earth presence to each other. The great value of this approach is that we have in the earth an extra-human referent for all human affairs, a controlling referent that is a universal concern for every human activity. Whether in Asia or America or the South Sea Islands, the earth is the larger context of survival.

All human professions, institutions, and activities must be integral with the earth as the primary self-propagating, self-nourishing, self-educating, self-governing, self-healing and self-fulfilling community. To integrate our human activities within this context is our way into the future.

### *QUESTIONS FOR DISCUSSION AND REFLECTION*

1. Our industrial society has provided us with the highest standard of living in the history of humans. What reasons do Berry and others have for critiquing industrial society *fundamentally*?

2. If one accepts Berry's critique, one faces a whole new way of looking at what we term "progress." What vision of progress most appeals to you?

3. Berry speaks of "visionary" and "mythic" as important for our industrial economy. Can you cite any examples of what he means from your experience?

4. Berry argues that Christianity has a great deal to do with our present crisis. He cites various historical reasons. Do you think it is plausible to "blame" Christianity, or do all cultures devastate the natural world to the same degree?

5. If you take seriously the need for radical change in our culture, what do you think should be done by your community, your local region, or your country to begin to deal with the problem?

6. Can the new story of the universe truly energize people to deal with this new crisis in a way that traditional religions, by themselves, cannot? In what ways?

## 2

# THE EARTH:

## A New Context for Religious Unity

Thomas Berry

Presently Catholics are less than twenty percent of the human community. As a people, we no longer exist in a completely unified society as in the European Middle Ages. Rather, the once catholic world is broken up into numerous Christian groups. Also, by immigration and by conversion, a large number of non-Christian religious groups and individuals has developed throughout the western world. Abroad, the church is scattered throughout the various continents and peoples of the world. In addition, many large and small Catholic communities exist in authoritarian nations dominated by a militant secularism.

A more tolerant secularism is pervasive throughout the entire world and tends to dominate life ideals as well as national and international institutions. In view of all this, the religious situation of the church of the Catholic community has become exceedingly complex. The need for a more formal consideration of the church in its relation to the religious and non-religious world about us is clear.

The Second Vatican Council gave much attention to this subject in several of its documents: *The Church in the Modern World; Ecumenism; Missions;* and *On the Church in Its Relation to Non-Christian Religions.* We are concerned here with this last document.

It is a brief, incidental declaration of the Council which evidently felt itself forced to say something significant without venturing very far into a subject too sensitive for any thorough treatment at the time.

What was said, with extreme caution, is, however, worthy of serious consideration since now more than ever the parish priest as well as any teacher of religion is constantly faced with questions concerning other religious traditions. Many of our best people, especially our young people with strong religious attractions, often find the spiritual disciplines and meditation practices of Asian religions or of the Sufi orders much more helpful than what is available to them in our church preaching or even in our religious communities.

### *FUNDAMENTAL RENEWAL*

All of these non-Christian traditions, including the spiritual traditions of native American peoples, are in a vigorous phase of development in what might be considered a fundamentalist type of renewal. This includes intellectual self-understanding, spiritual practice, and social presence. These fundamentalisms as well as Christian fundamentalist movements are in reaction to the trivialization that has eventuated from so many efforts at liberal adaptations to the modern world.

This effort at fundamentalist integrity in various traditions exists along with increased study of these traditions and of influence from these traditions. Religiously the Christian west has probably been more influenced by these other traditions than they have been influenced by it, although both have been overwhelmingly influenced by the modern secular world.

In this situation the church has sought guidance for its actions from its own biblical, patristic, and theological traditions. In the document *Nostra Aetate* (*In Our Age*), there are the usual references to biblical passages asserting the divine concern for all people. In a rather guarded manner the statement is made that other religions often reflect "a ray of that truth which enlightens all men." This is followed immediately by a reference to Christ as "the way, the truth, and the life in whom men find the fullness of religious life, and in whom God has reconciled all things to Himself" (2 Cor. 5:18-19). This assertion betrays a certain anxiety lest admission of any authentic revelatory experience outside the Christian tradition lead to a diminution of the Christian claim to the integral revelation of the divine to the human community.

This way of dealing with the inter-religious issue has its own validity though it is not particularly new in its overall vision. Throughout past centuries, substantially the same thing has already been said. But now, through our religious studies, we are able to identify in greater detail the ways other religions reflect not only "a ray" of the divine light but even floods of light illumining the entire religious life of the human community. The Council did clarify in a few succinct statements the Catholic view that revelation is not absent from the non-Christian world; that all that is spiritually good should be encouraged and developed in native traditions; and that the peoples of earth comprise a single community with a single origin and a single destiny and that a single providence communicates its saving design to all peoples.

Behind these statements are the basic questions of unity and diversity and how differences relate to each other in some meaningful unity. In the attitude of the Council, and in the Christian tradition from the beginning, we find a powerful sense of unity, a suspicion if not abhorrence of diversity in religious concerns. This attitude originates in an overwhelming sense of the oneness of a personal transcendent divine creator. Originally a tribal deity, Yahweh was elevated to singular status over against all other deities. Yahweh absorbs into himself all that divine power generally experienced as diffused throughout the universe and articulated on the planet earth in the manifold phenomena of the natural world. Associated with this deity is a singular people bound by a covenant expressed in written form which, in its later new covenant expression, is communicated to the peoples of earth through a divinely established hierarchical church.

This sense of an elect people as exclusive bearers of a universal salvation either originated in or was powerfully reinforced by the feeling that as a small people they had an important destiny that was constantly threatened by surrounding political powers. The more threatened this elect people felt, the more intensely they experienced their own significance as a people destined to be the instrument of divine rule over all the nations of earth.

While this attraction to an ambiguous political-religious rule by an elect people was spiritualized in the New Covenant period, the church did emerge with the sense of having an exclusive universal role in bringing about the spiritual well-being of a fallen world. The existing religions of the world were seen fundamentally as obstacles, although the Holy Spirit never failed to be present to well-disposed individuals.

### *DIVERSITY NOT APPRECIATED*

While a multitude of scriptural, patristic, and theological quotes can be given in support of a larger overall vision of the salvific process, these are generally extenuations or remedies for a fundamentally undesirable situation. Worst of all is the lack of appreciation of that diversity which Thomas Aquinas designates as the "perfection of the universe." By definition the universe is diversification caught up in a complex of functional relationships. Obviously the diverse part in its particularity cannot itself constitute the principle of unity, although each part articulates the whole in some unique fashion. The perfection, however, is in the whole; not in the part as such.

Here it would be of some help to quote from that same article of Aquinas from the *Summa Theologica* (Part I, Question 47, article 1). In this question dealing with the distinction of things, Thomas remarks in a concluding statement that the multitude of things comes from the first agent who is God.

> For He brought things into being in order that His goodness might be communicated to creatures, and be represented by them; and because His goodness could not be adequately represented by one creature alone, He produced many and diverse creatures, that what was wanting to one in the representation of the divine goodness might be supplied by another. For goodness, which in God is simple and uniform, in creatures is manifold and divided; and hence the whole universe together participates the divine goodness more perfectly, and repre-

sents it better than any single creature whatever.

This law of diversity holds not only for the other areas of being and of action but also for the religious life of the human community, for revelation, belief, spiritual disciplines, and sacramental forms. If there is revelation it will not be singular but differentiated. If there is grace it will be differentiated in its expression. If there are spiritual disciplines or sacraments or sacred communities they will be differentiated. The greater the differentiation, the greater the perfection of the whole since perfection is in the interacting diversity; the extent of the diversity is the measure of the perfection.

When the religious traditions are seen in their relations to each other, the full tapestry of the revelatory experience can be observed. Each articulated experience is shared by the others. In the fabric of the whole, the divine reveals itself most fully. This requires a threefold sequence of emphasis in the various traditions. First there is the primordial experience expressed in an oral or written form, the scriptural period. This takes place by an isolation process. Secondly, there is the deepening of the tradition, a patristic period, when the implications of the original experience are elaborated in contact with the larger life process. Thirdly, there is the period of expansion, of interaction with other traditions in their more evolved phase. These three phases are not mutually exclusive since some interaction of traditions is present from the beginning both in assisting positively and in providing a polarity of oppositions.

The importance of the isolation and interior development phases can be seen in each of the major traditions that have so powerfully influenced the religious life of the human community. In India, for instance, its profound mystical and metaphysical developments of the early first millennium B.C. required a special type of psychic intensity.

In China the focus of attention was much more cosmological. The divine was understood as Shang-ti, as T'ien, as the great mystery presenting itself in the vast cosmological cycles in which the human was also a functional presence. The divine, the natural, and the human were thus present to each other in the grand sacrificial reality of the universe itself and in its rhythmic pulsations.

So too with the Japanese and their sense of the aesthetic expression of the numinous in the natural world. Their cultivation of spiritual simplicity and spontaneity is unique in the human community. So with the presence of the Great Spirit throughout the natural world and in the human heart in the American Indian traditions. In every case these ultimate orientations toward reality and value originate in an interior depth so awesome that the experience is perceived as coming from a trans-phenomenal source, as revelatory of the ultimate mystery whence all things emerge into being.

### *A NEED FOR THE DIVINE*

To maintain that these experiences are simply the consequences of natural reason and not valid revelatory experiences communicated by the divine would negate the sacred character of those most profound of all experiences that have shaped the psychic structure of the human. Certainly India is absolutely clear on the need for the divine to reveal itself in some active, positive manner if humans are to know the divine as such. All the great Asian traditions have explicit statements on the need for grace, for a special mode of divine activity to effect the spiritual fulfillment of the human.

Through the support of those sacred experiences not only Asian peoples but peoples everywhere have been able to attain sublime spiritual insight and to endure and even to exult in a life of much struggle and pain along with its ecstatic delight. The language, customs, and artistic styles of peoples everywhere all give expression to this presence of the divine. Each of these revelatory traditions in its own mode reaches unequaled levels of religious experience.

All of these traditions were substantially complete in their earlier expression. Hinduism in its proper line of development will not likely go beyond its expression in the Hymns, the Upanishad, the Epics, the Bhagavadgita. Buddhism in its proper line of development will hardly go beyond its expression in the early dialogues, the Dhammapada, the Sutta Nipata, The Lotus Sutra, the verses of Nagarjuna, the Vimalakirti Story. So with the Confucian classics, the Four Books, the Tao Te Ching, the Writings of Chuang-tzu. These are all full and perfect in their own context although each is a vital expanding process with an ever-renewing series of transformation through the centuries. So, too, the Christian revelatory experience is full and complete in its own proper mode. More than any of the other traditions, the Christian experiences revelation through a sequence of historical events as opposed to the metaphysical, the mystical, the cosmological, the psychological, or the immediate experience of the numinous dimension of the natural world.

While all these developments took place in limited geographical areas and in limited modes of consciousness, each of these revelatory experiences were considered as comprehensive interpretations of the universe and as effective guides for individuals and communities in attaining their divinely determined destiny. Although this sense of completeness existed in the various traditions, they have generally been open traditions, willing to interact creatively with other traditions, both listening and speaking.

The deepest value of each tradition, however, has been in its own distinctive insights. It would not have been possible for India's experience of the transphenomenal world to be developed in its full intensity simultaneously with the biblical experience of the revelatory aspect of historical events. Nor could China have developed its insight into the mysterious Tao of the physical universe simultaneously with the high metaphysics of India. Nor would the experience of the Great Spirit manifested in the natural world, which is enjoyed by the indigenous peoples of the North American continent, be compatible with a simultaneous experience of the Buddhist doctrine of Emptiness or the neo-platonic doctrine of the Logos.

None of these experiences are rivals of the others. Each needs a certain isolation from the others for its own inner development. Each is supreme in its own order. Each is destined for universal diffusion throughout the human community. Each is needed by the others to constitute the perfection of the revelatory experience. Furthermore, each carries the whole within itself, since part and whole have a reciprocal relationship. Each has its microphase-initiated institutional adherents and its macrophase or universal presence throughout the human community. We are now living in the macrophase period of development of most religious traditions, the period of extensive influence without formal initation. This is possible because none of the traditions do precisely what the others do. Each has, as it were, its own area of spiritual consciousness. The traditions are, as it were, dimensions of each other.

### *THE REVOLUTIONARY EXPERIENCE*

In this perspective, the Council document on Revelation should be entitled "Christian Revelation" or "Biblical Revelation," since it does not deal with the revelatory experience as such. Also, any reference to Christian revelation as the "fullness" of revelation must consider the precise sense in which this term is being used since the Scriptures and the Christian tradition themselves indicate that there exists a revelatory communication of the divine throughout the human community. If this be true, then the fullness of the revelatory experience is in the larger range of these highly differentiated experiences.

The difficulty in statements about the Holy Spirit being present to all peoples is the implication that the Holy Spirit is communicating to others the same thing that has been revealed in the biblical-Christian traditions, only less clearly or less completely. The difference is seen as quantitative rather than qualitative. If the difference is qualitative, then according to traditional views, these traditions do not qualify as "revelation" but as some "natural" mode of knowing. If the difference is quantitative, then according to traditional views Christian superiority can be stated in terms of fullness or completeness.

The first statement is where the difficulty might be located. I am proposing a qualitative difference within the authentic revelatory process itself, a difference that cannot be resolved in terms of fullness or completeness but only by mutual presence of highly differentiated traditions. What is to be avoided is any monoform tendency in the meeting of religions. What is to be sought is a mutually enhancing meeting of qualitatively differentiated religions in which both the microphase and the macrophase expression of each religion would benefit.

The attraction toward "one flock and one shepherd" is a seductive attraction that easily leads to sterility in the religious process. It is a kind of escape, a rejection of reality, an attitude that diversity should not be, that it is a hindrance to human well-being and to the salvific process, that it must be ended as quickly as possible. This attitude has led, in the academic world, to such an aversion toward other religions that nowhere on the North American continent is there a Catholic university where professional studies on these other religions can be done. Nor in our theological centers is there adequate concern for these other religions.

Because of this attitude Catholic interaction with other religions is diminished in its efficacy. The challenge is avoided; the tension needed for creativity is lacking since the basic principle of life and movement is unbalance rather than balance, asymmetry rather than symmetry. Diversity is enrichment. For the biblical concept to be the universal concept of deity to the elimination of Shiva and Vishnu, of Kuan-yin and Amida, of Shang-ti and T'ien, of Orenda, Wakan-tanka and the Manitou, would be to impoverish the concept of deity. For the Bible to be the only Scripture to the elimination of the Vedic Hymns, the Upanishads, the Bhagavadgita of India; to the elimination of the Koran of Islam, the Lotus Sutra of Buddhism, the Sacred Books of China, would constrict rather than expand divine-human communication.

For any situation the ideal is the greatest tension that the situation can bear creatively. Although every archetypal model needs multiple realizations, the sacred, more than any other element of reality, needs variety in its modes of expression.

The difficulties experienced by Christians in accepting the variety of religious traditions can be resolved:

1) By distinguishing the microphase membership and macrophase influence of all religious traditions.

2) By identifying the unique communication by Christian revelation in both modes of its expression.

3) By recognizing the qualitative differences in religions and fostering these differences.

4) By identifying the dynamics of inter-religious relations.

5) By fostering a sense of the New Story of the universe as the context for understanding the diversity and unity of religions.

Our discussion so far has been within the general context of traditional modes of thinking with no significant reference to the vast changes in our modern way of experiencing the universe, the human community and the modes of human consciousness. Yet these are powerful determinants in all our religious as well as in all our cultural developments.

## *A NEW CONTEXT*

To talk about religious traditions simply out of their own inner processes is to ignore this larger context of interpretation. Accepting this new context is difficult for us because of the prevailing secular culture, the materialist view of the universe and rationalistic modes of contemporary thinking. Because these attitudes have been so antagonistic to religion we can hardly believe that the long course of scientific meditation on the universe has finally established the emergent universe itself as a spiritual as well as a physical process and the context for a new mode of religious understanding. We might describe it as a meta-religious context for a comprehensive view of the entire complex of religions.

We have rewritten *The City of God* of Saint Augustine, not this time as the story of two cities seeking the extinction of each other, but as the story of an immense cosmic process, both spiritual and physical from the beginning, articulating itself in ever greater variety and complexity until it has come to a certain fullness of expression in human consciousness. Within human intelligence the creative process attains a capacity for self-awareness and for a human inter-commmunion with the numinous mystery present throughout this process. This inter-communion, as a revelatory presence of the divine, takes place throughout the human community in the diversity of its manifestations. From these primordial indigenous experiences have come the diverse Scriptures of the world, the various forms of worship, the variety of spiritual disciplines.

I suggest this context of interpretation for the diversity, identity, and inter-communication of religions. It might be considered as a cosmological-historical approach over against the traditional theological, sociological, or psychological approaches to the subject. This cosmological approach accords with the basic statement of Thomas Aquinas concerning the cosmic community as the "perfection of the universe," as the supreme reality which "participates the divine goodness more perfectly, and represents it better than any single creature whatever." It also accords with the view of Teilhard de Chardin that "man is a cosmic phenomenon, not *primarily* an aesthetic, moral, or religious one" (Pierre Leroy, *Letters From My Friend: Teilhard de Chardin*, [New York: Paulist, 1980]).

If the human can only be understood within the universe process, then every aspect of the human, including the religious dimension of the human, is involved. But whether we begin with Aquinas or with Teilhard, the universe is the primary religious reality, although this religious dimension, as well as its psychic dimension, is fully articulated only in the human. In no instance, however, is the human in any of our activities functioning simply by itself, independent or isolated from the universe process that brought us into being, sustains us, enlightens, and sanctifies us.

## *THE UNIVERSE INSTRUCTS*

If humans have learned anything about the divine, the natural or the human, it is through the instruction received from the universe around us. Any human activity must be seen primarily as an activity of the universe and only secondarily as an activity of the individual. In this manner it is clear that the universe as such is the primary religious reality, the primary sacred community, the primary revelation of the divine, the primary subject of incarnation, the primary unit of redemption, the primary referent in any discussion of reality or of value. For the first time the entire human community has, in this story, a single creation or origin myth. Although it is known by scientific observation, this story also functions as myth. In a special manner this story is the over-arching context for any movement toward one creative interaction of peoples or cultures or religions. For the first time we can tell the universe story, the earth story, the human story, the religion story, the Christian story, and the church story as a single comprehensive narrative.

The choice, however, of the Council was to establish the biblical-redemption story rather than the modern creation story as its context of understanding. In doing so it set aside its own most powerful instrument for dealing with the church, revelation, the modern world, missions, and its relation with non-Christian religions. It will undoubtedly be a long time before such a transition in our thinking will take place.

We might propose, however, that until this new context for understanding is accepted, the unique role of Christianity and of the Catholic Church will not attain its full efficacy. If Saint John and Saint Paul could think of the Christ form of the universe, if Aquinas could say that the whole universe together participates in the divine goodness more perfectly and represents it better than any single creature whatever, and if Teilhard could insist that the human gives to the entire cosmos its most sublime mode of being, then it should not be difficult to accept the universe itself as the primordial sacred community, the macrophase mode of every religious tradition, the context in which the divine reality is revealed to itself in that diversity which in a special manner is "the perfection of the universe."

## *QUESTIONS FOR DISCUSSION AND REFLECTION*

1. Thomas Berry focuses on the question of unity and diversity, which raises a problem since Christianity has always emphasized the conversion of all people to Christ. Berry believes that revelation is wider than Christianity because there are different facets of the divine revealed in other religions, all of them necessary. Is this compatible with your sense of being a Christian? Why or why not?

2. If we believe in the diversity of religious traditions and in the perfection of that diversity, would this be reducing all religions to the same thing? Why or why not?

3. "Each is supreme in its own order." Is this vision of relig-

ious diversity attractive to you or simply confusing? What difficulties or insights does it bring you?

4. Berry stresses creative tension and diversity rather than harmony, community, and symmetry. In what ways can you resonate with this vision? What bothers you about it?

5. How do you understand this quote from Teilhard: "Man is a cosmic phenomenon, not primarily an aesthetic, moral, or religious one"?

# 3

# *THE EARTH STORY:*

## *Where Does the Bible Fit In?*

Donald Senior, C.P.

*Donald Senior, C.P., of Catholic Theological Union in Chicago, and author of many biblical studies, expresses a profound gratitude to Thomas Berry for the sweep of his perspective. "Perhaps," he says, "the most important thing I can say is that I find Berry's statement of the issues facing humanity most compelling. . . . While most commentators point to 'micro' elements missing from analysis of our plight, Berry's perspective sweeps wide and deep, fitting humanity into its proper environment."*

*In spite of Berry's critique of the biblical tradition, Senior feels that there is a fundamental continuity between his position and many biblical themes. Secondly, he realizes Berry is asking for conversion, and "For Christians, at least, there will be little possibility for such conversion if that means repudiation of the biblical story." Senior also challenges Berry to clarify his ideas on sin and evil.*

I would like to offer here a brief response, from a biblical perspective, to some of the writings of Thomas Berry. At the outset I want to state my long-term appreciation for Berry and his important work. It is a privilege and a pleasure to be able to interact with him through this book. The immediate and exclusive focus for my comments are his two papers, "The Earth: A New Context for Religious Unity" and "Economics: Its Effect on the Life Systems of the Earth." These writings may not represent the full scope of Thomas Berry's remarkable range of interests, but they strike me as good samples of his current concerns and theological style.

### *BASIC APPRECIATION OF MAJOR THEMES*

Even though these papers address two different areas (ecumenism and economics), they exemplify common concerns and a common approach. I would like to cite what I consider the major elements in each of them:

1) First of all, seeking communion among the various religious traditions of the world or trying to create an equitable and efficient economic system both demand an overarching "vision." As Berry notes in "Economics,": "However rational modern economics might be, the dynamic of economics is visionary." In "The Earth" Berry refers to this as developing a "new story," which would enable the religious traditions of the world to see themselves in a new relationship to the universe.

2) Such a vision and the commitments and actions resulting from it must be grounded in a comprehensive sense of reality. Too often the guiding vision or story of a community has excluded major components from its purview, resulting in a false and exclusive sense of identity. In the case of economics, failure to conceive of the earth as a planetary community and as a part of the universe puts humanity on the road to suicidal destruction of the life-supporting systems of the world.

3) Diverse elements are multiple reflections of the One, therefore, the diversity and unique identity of each component in the human and universal community is to be recognized and enhanced. This is particularly clear in the case of ecumenism. The oneness of God is not threatened by diversity of religious intuitions but properly revealed. Therefore uniformity cannot be the vision or endpoint for the quest of oneness in the religious sphere.

4) Another principle, flowing from the previous, is that at the core of all human endeavor must be the search for community that

encompasses the full range of reality. Human community must not be sectarian or exclusive but must be open to all. As Berry notes: "...the community of all the components of the planet earth is primary in the divine intention."

5) In all of the areas cited above, Berry is convinced that Christian tradition has not only failed to adequately respond to the present urgent need for a "new story" or a new "vision," but has also been responsible for much of the false and harmful ideology that has brought humanity to the brink of self-destruction. As I will comment below, the biblical tradition seems to have been one of the prime offenders.

Perhaps the most important thing I can say is that I find Berry's statement of the issues facing humanity most compelling. In economics and ecumenism—and in a host of other urgent human issues—we are, it seems to me, dogged by collective myopia. While most commentators point to "micro" elements missing from analysis of our plight, Berry's perspective sweeps wide and deep, fitting humanity into its proper environment.

Without such realism—a realism unfortunately tagged as "unrealistic" in many quarters—neither the true proportions of our problems, nor the potential for our solutions, can be found. And Berry is also correct in pointing out that all human endeavors, however mechanistic they might seem, work out of a "vision." Attention to that vision is essential. And, finally, the search for an earth-wide community, from which no one or no thing is excluded, is surely an eloquent statement of a vision ratified by the deepest and truest instincts of humanity.

## *POINTS OF CRITIQUE*

In all these things and more I find myself in agreement with Thomas Berry and grateful for his eloquent statement of them. However, this book would be far too placid if all the contributors dwelled on vast areas of agreement. So let me give more time to those aspects of Berry's two papers with which I would differ or at least question.

I should point out that I believe the Bible would be in strong agreement with the principles found in Berry's two papers. The Bible itself is the articulation of a "vision" or "story" enabling the people of Israel and the early Christians to take a dynamic approach to their worlds. And a case could be made that "realism" is a primary biblical virtue. While capable of lyrical poetry the biblical literature is not utopian. It is drenched in human experience and pushes its reflections to every corner of its perceived world. And one could not deny that a vivid sense of community is at the heart of the biblical dream.

In citing such agreements, however, I may already find myself at odds with Thomas Berry's viewpoint, because he seems to find little support for his own analysis in the Bible. And, further, he believes that the Bible is responsible for some of the problems that stand in the way of achieving authentic community. I believe that this aspect of these papers needs to be reassesed for two reasons: 1) the statement of the biblical traditions on some major points strikes me as inadequate; 2) important biblical support for the vision or story Berry suggests is overlooked.

My attempt to defend the biblical viewpoint is, of course, predictable. But I hope what I have to say will be seen as something more than attempting to fine-tune statements of biblical theology or as a defensive reaction to criticisms of the beloved good book. It struck me that if Berry's analysis of the human plight is correct and if, in fact, humanity needs to so thoroughly revise and expand its vision (or create a new story) then what is being called for is "conversion"—in the most serious sense of that word. However, for Christians at least, there will be little possibility for such conversion if that means repudiation of the biblical story. Only in continuity with our sacred story can the energy be found to effect radical change in perspective, and the foundation of our Judaeo-Christian story is biblical.

## *MORE ROOM FOR DIVERSITY*

First of all, I think there is much more room for diversity in the biblical picture than Berry seems to find. Even the Bible's most exclusive categories such as "election" (the prime example used in

Berry's paper on ecumenism) are not absolute. It is too simple a portrayal to see election giving the religious community of Israel absolute claims to truth or land or destiny, at the expense of others.

As Walter Brueggeman pointed out in his study, *The Land* (Philadelphia: Fortress, 1982), Israel's claims as an elect people were always tempered by the reminder that they themselves did not own the land and had not gained it through their own efforts, that dwelling in the land meant responsibility for the land, and that recognition of the land as gift called for greater fidelity to the Torah, especially regarding the claims of justice on behalf of the poor and the sojourner.

This theology is strongly to the fore in the book of Joshua and Deuteronomy as Israel is pictured standing on the brink of claiming its "promised land." It is a caricature of the biblical tradition to see election in any of its expressions (land, temple, monarchy, etc.) as giving the warrant for exploitation of its elect status at the expense of others or of the land itself. The *herem* or ban that became part of Israel's theology has the limited function of illustrating the purification and commitment needed for fidelity to God; it is not the totality of the biblical viewpoint concerning the "outsider."

In our book, *Biblical Foundations for Mission* (Maryknoll: Orbis Books, 1983), Carroll Stuhlmueller and I tried to demonstrate that pluralism and diversity have always been part of the biblical picture, even in times of strong sectarian pressures. At the same time that the post-exilic crisis produced the sectarian viewpoint of Ezra and Nehemiah, you also have more universalistic strains of Jonah and Trito-Isaiah. At the same time the book of Revelation calls for prophetic withdrawal from any compromise with the demonic Roman system, the author of 1 Peter calls for good citizenship and discriminating participation in all "created structures" (1 Peter 2:13), including civil society.

Is it accurate for Berry to say that the biblical writings are produced by an "isolation process"? Some are, undoubtedly, but much of the theology of the New Testament (and a good deal of the Old) was stimulated by contact with peoples outside of the community of Israel or the Jewish-Christian community. Some of the most vital moments in biblical history came not in isolation but in interaction with its surrounding environment. Much of the purpose of Luke-Acts, for example, is to validate the movement of the community from a Palestinian context to a Hellenistic context.

Paul's most eloquent theological moments come in defense of his inclusion of Gentiles against those who would have restricted their freedom (e.g., Galatians) and, equally revealing, his defense of a non-Christian Judaism in Romans 9-11. Similar cases could be made for the rest of the Gospels and much of the other New Testament books. A centipetal dynamic is by no means the only biblical dynamic.

It is true, as Berry points out, that the Bible does not take a positive view of other "religions." But here, too, I think we must look carefully at our statement of the issue. Is it not anachronistic for us to see the biblical peoples thinking in terms of "religions" or "religious systems"? I think this is a thoroughly modern consciousness. Most biblical reflection concerning non-Jews is strongly relational, not abstract or analytical or systemic. That is, most of the biblical drama is played out, not in comparing "religions," but in determining allegiances.

Israel dealt with Egyptians or Canaanites or Amorites just as the early Christian community dealt with Greeks in Athens or Thessalonica, or Romans in Corinth and Rome itself. There are few if any biblical texts that deal explicitly with "Canaanite religion" or "the cult of Isis." Therefore an analysis of the biblical viewpoint towards non-Christian religions, for example, would have to consider biblical attitudes to the Gentile. These attitudes are not monolithic in either testament. The "Gentile" could be seen as enemy or instrument of God (e.g., Cyrus); as object of conversion or as exemplar of virtue (e.g., Cornelius). There is, evidently, a conviction that worship of Yahweh or faith in Christ is ultimate and decisive truth but this conviction was not systematically compared with the truth claims of other religions as such.

In short, the biblical viewpoint itself is pluralistic and there is much more interaction with other peoples and values in the biblical tradition than is often acknowledged. As Ernst Kasemann noted in referring to the New Testament canon, the Scriptures are the source not only of our unity but of our diversity.

### *THE "IMAGE OF GOD" PROBLEM*

On two counts, Berry finds the biblical image of God a contributing cause to the exploitation of the earth and the fragmentation of the human community. First of all the "biblical commitment to a transcendent personal monotheistic concept of deity with severe prohibition against worship of divinity resident in nature" had the effect of "despiritualizing" and "desacralizing" the natural world. Secondly, the biblical emphasis on redemption rather than creation also leads to an exploitive view of the natural world.

On both counts I find these statements of the biblical tradition too sweeping and too focused on certain biblical traditions to the exclusion of others. While surely the Bible conceives of Yahweh as transcendent, personal, and making exclusive claims for allegiance, I am not sure that the other side of the equation, a consequent desacralization of nature, holds true. The biblical critique of Canaanite religion was not primarily or specifically because of its nature orientation but because biblical peoples viewed such rites as expressions of allegiance to an "alien" God and were often offensive to Jewish sensibilities (e.g., sexual ritualizations). And determining the biblical reactions to "pagan" rituals must be nuanced; in fact, Israel adapted many of the nature rituals of the Canaanites: Passover, Pentecost, Sukkoth, to name three of the most important Jewish festivals.

Biblical traditions as varied as the Pentateuch, the Wisdom literature, and the pauline letters (e.g., Romans 1:19-23) viewed the earth as created through God's word and as an effective revelation of God. Even the non-Jew was expected to find God in the structures of nature. Matthew's Gospel does not hesitate to portray the magi as finding the Messiah through their search of the stars. Symbols of the good earth, renewed and abundant, figure prominently in prophetic visions of the end time (e.g., Isaiah 25, 65, etc.).

In short, I find little biblical evidence that allegiance to Yahweh led to a less reverent view of the sacred or spiritual nature of the earth. (Just as I wonder if there is any real evidence today that peoples who adhere to some form of nature religion are less inclined to pollute the earth once they have the technology to do so.)

Secondly, one must concede that redemption is a more prevalent category than creation in the biblical tradition. But are biblical statements about redemption warrants for a mythology of unlimited progress? Does not such a mythology result precisely from ignoring essential elements of the biblical notion of redemption: e.g., land and life as gifts of God and not, ultimately, within the dominion of the human; the inclination of humanity to sin; the need for ongoing conversion; the call to justice, etc.? In the theology of Deutero-Isaiah, creation and redemption are merged into a single category; creation is conceived not as a past, "paradise-lost" phenomenon, but as an ongoing process by which God shapes humanity and its world (similar, I believe, to the emphasis Berry calls for: "Creation...must now be experienced as the emergence of the universe as a psychic-spiritual as well as a material-physical reality from the beginning").

In short, I do not find compelling evidence in the Bible that an emphasis on God as redeemer implied a desacralization of the natural world or encouraged an exploitive attitude toward the earth. Instead, I believe the biblical viewpoint on almost every count points to a full integration of the physical and the spiritual, of the heavenly and the earthly. If a scapegoat must be found for the presence of this alienation in western culture, I would look to the worldview of Greek thought and to the urbanized context of later European experience.

### *POWERFUL BIBLICAL TRADITIONS*

Finally, I would suggest that some biblical traditions that might be powerful foundations for the theological vision Berry portrays seem to be untapped. For example, how does the Christian (and Jewish) notion of resurrection fit into all this? It is not mentioned in Berry's two papers; yet it strikes me that resurrection, conceived as a metahistorical transformation and integration of matter and spirit, is the ultimate foundation for "planetary socialism." Resurrection keeps the Christian earthbound in the most authentic way and is a deterrent to placing alienation between body and spirit.

Another helpful motif that offers important traditional grounding for the scope of Berry's theology is the "cosmic" perspective of some biblical traditions. Berry alludes briefly to this in referring to the cosmic Christology of John's Gospel and the letter to Ephesians. In fact, the cosmic perspective of John and Ephesians are part of a broad biblical perspective that is grounded in the Wisdom tradition (cf. the cosmic wisdom poetry of Sirach, Wisdom, Proverbs) and influenced other New Testament writings such as Matthew (Ch. 11) and Paul (cf. Colossians, Philippians).

Judaism has poetically conceived of the universe as created in the image of God's wisdom and this poetic-metaphysics became the framework in which early Christianity could see Christ's (and therefore humanity's) relationship to the cosmos as well (cf. J. Dunn, *Christology in the Making* [Philadelphia: Westminster, 1980, pp. 163-212]; M. Hengel, *The Son of God* [Philadelphia: Fortress Press, 1976]).

One final point that is more question than comment. Although in these two papers Berry eloquently describes humanity's contemporary plight, his notion of evil is not clear to me. The Bible, of course, wrestles constantly with the mystery of evil. Is the evil we experience through pollution of the earth or in the failure to appreciate the religious insights of other traditions due solely to false ideology? In other words, with proper *understanding* would our problems be solved? Or is it a problem not only of perception but of *power*, i.e., does humanity have the capacity to be open to truth and to community? The biblical view leans more to the second way of framing the question of sin and conversion. Theology that is creation-centered has been accused of stumbling over the question of evil. It would be interesting to hear Berry's view of this.

I conclude my comments where I began: I am grateful for Thomas Berry's penetrating theological vision and have been taught and inspired by it. I am convinced that his analysis would only be more effective if biblical traditions were seen less as culprit and more as ally in the task of creating a global, inclusive community. As Berry notes, the relation of the Bible to Christianity is not identical to that of the Koran to Islam, for example.

Roman Catholic Christianity, at least, has always maintained that biblical revelation is linked to human experience and that the insights of biblical revelation can be properly interpreted only in light of the Spirit's ongoing work in the world. The need to develop a new story is therefore not really alien to the Christian vision (even though it may be resisted). But at the same time, as suggested in Matthew's Gospel, the Christian scribe who tells the new story must bring from the treasure house things that are old, as well as new.

## *QUESTIONS FOR DISCUSSION AND REFLECTION*

1. Senior's position might be summarized as a justification of the biblical vision of "human beings in community and that community linked to the land." He finds this very compatible with Berry's vision, but he does not think the biblical tradition is the cause of our western disregard of the natural world. He suggests that Greek thought, and later, the rise of capitalism, are key in this development. What do you think?

2. Thomas Berry is calling for conversion, but Senior states that no conversion can take place unless it is linked to one's history or tradition. What link can you make with the Christian tradition in the thought of Thomas Berry?

3. Do you think there is a general insensitivity to nature in conciliar statements, anthologies of Christian literature, liturgies, and encyclicals? Why might this be?

4. Berry and Senior agree that historically revelation is always being used in the context of biblical revelation, and that the need today is to listen to the revelation of the universe. What do you think they mean by this distinction?

5. Senior writes: "I am convinced that his [Thomas Berry's] analysis would only be more effective if biblical traditions were seen less as culprit and more as ally in the task of creating a global, inclusive community." In what ways might the Bible be an ally?

4

# *THE GRAND VISION:*

## *It Needs Social Action*

Gregory Baum

*Gregory Baum, formerly professor at St. Michael's College, Toronto, and currently of the Faculty of Religious Studies at McGill University, Montreal, comments on Berry's work from the viewpoint of a theologian and political activist. While he shares Berry's concerns about global issues, he disagrees with what he identifies as a romantic rejection of the Enlightenment with its ideals of rational control and scientific freedom. The original enlightenment project, he maintains, must be distinguished from the scientific enlightenment that eventually triumphed. What follows for Baum is a commitment to socialism, since socialism is essential if society is to limit its growth and develop industry in keeping with the requirements of ecology. He contends that Berry's analysis detaches people from identifying with the socialist movement. He also takes issue with Berry's distance from Christo-centered theology.*

Thomas Berry's radical proposals deserve admiration. They go against the narrow limits of traditional theology and dominant political science. For a priest to adopt such radical views demands courage. Moreover, Thomas Berry focuses on two important global issues that may well define a turning point in the world history, issues that do not receive the attention they command: a) the ecological concern for the earth's surface and b) the cooperation of the world religions to protect global life.

While I admire Berry's proposals, I do not agree with them as they stand. I wish to mold them a little. Here are my reasons.

### *WE STILL NEED ENLIGHTENMENT*

Because of the impasse to which modern technological and bureaucratic control has led society, certain social critics, including Thomas Berry, have suggested that the entire modern project, the enlightenment tradition, has been a giant mistake. They claim that the modern project was based on the ideals of scientific mastery, progress, rational control, and emancipation. What the modern project overlooked was humanity's rootedness to nature. The human task, these thinkers argue, is not to dominate nature but to be reconciled with it. What we have to do, they propose, is to reject the enlightenment and create an alternative paradigm of rootedness and reconciliation.

I have little sympathy for the romantic rejection of the enlightenment. With European and North American "Political Theology," I have been greatly impressed by the analysis of the Frankfurt School (Horkheimer, Adorno), according to which the contemporary scientific enlightenment is indeed a death-dealing power: it is the problem, not part of the solution. In our day rationality has become equated with technological or instrumental reason: reason is here concerned with means, no longer with ends; reason has become positivistic and anti-humanist.

Yet the original enlightenment project was different: Here reason embraced scientific rationality as well as practical reason, i.e., reason as normative guide to a truly human life. The original enlightenment regarded reason as the faculty by which people were meant to discover their true vocation and free themselves from the oppression inflicted on them. Reason had here an ethical, critical, practical dimension.

Yet in the course of the last two centuries, the enlightenment tradition has become its own enemy. Reason has collapsed into purely technological or instrumental rationality. We have repressed the critical, ethical dimension of reason. In the view of the Frankfurt School, the response to this dilemma is not the romantic repudiation of the enlightenment, but the critical negation of its present instrumental one-sidedness and the retrieval of practical, ethical reason.

This dialectical negation of the enlightenment has appealed to Catholic theologians in particular. The retrieval of practical reason corresponds to the traditional Catholic conviction that reason is a faculty that discerns and grasps the substance of human existence. While the reason of common sense is distorted by the cultural mainstream and hence in need of redemption, a deeper 'ratio,' available through a quest that includes conversion, is a reliable guide to the ethical life. Following "Political Theology," I agree with Berry's critique of the scientific, technological, and bureaucratic world, but against Berry I refuse to reject the entire modern project. We are still in need of enlightenment and emancipation. We need to retrieve practical reason. What is demanded here is ethical conversion, which becomes available to us as we engage in action, as we act in solidarity with the victims of society. With "Political Theology," I argue for a certain affinity between the retrieved enlightenment and tradition and the liberating Christian message.

### *WE NEED POLITICAL INVOLVEMENT*

Related to this theoretical point is Berry's practical thrust. He has decided to define the ecological or "green" movement in opposition to all existing political orientations. This seems to me unrealistic. There is hope for us only if the green concerns can be integrated into a socialist project. While classical socialism was as growth-oriented as capitalism, this orientation was not intrinsic to socialism itself. On the contrary, if a society wants to limit its growth and develop an industry in keeping with the requirements of ecology, it must gain control over the means of production or at least over the use of these means. Berry's analysis detaches people from identifying with the socialist movement in their place. A different analysis would persuade people to join a socialist party and make it more sensitive to ecological issues.

Berry's essays reflect the alienation many Americans feel in regard to their institutions, their government, their political parties, their entire system. They do not see any action that could significantly change the orientation of their country. The U.S.A. has become an empire. Radicals in an empire easily turn to idealism, bold proposals, new dreams devoid of practical policies.

In Canada, we are not as alienated from our institutions. We are not an empire, our institutions are more amenable to political action. Many of us still hope that the political pressure on the government, supported by the churches, will result in the recognition of the land claims of the Native people. I understand there is no equivalent of this in the U.S.A. Or again, thanks to the political pressure of the government, backed by the churches, Canadian foreign policy toward Cuba and Central America is decidedly different from the American one. And while our support for Contadora and similar projects have not achieved a great deal, it would be rash to give up our political involvement in this field.

If we long for reconciliation with nature in Canada, we should support the farmers who try to save their tradition against the inroads made by agribusiness. Again, some churches have involved themselves in this. We have in Canada a political party that has inherited a home-grown socialism (the cooperative movement) and farmer's concerns, and while this party can be criticized from many points of view, in my opinion it deserves the support of all who love justice. This party is a source of hope for many critical Canadians. For this reason, radicals in Canada often try to formulate their proposals in a manner that promotes solidarity and cooperation in the same political movement. We have no green party, but we have a green movement in the NDP.

At this time, when the government has proposed free trade with the United States, many Canadians fear that this could be the beginning of our integration into the American economy, a process that would be accompanied by the decline of our bi-national culture, the dismantling of our social system, and eventually the collapse of our sovereignty. This has been an occasion for solidarity: the trade union movement, the NDP, the woman's movement, the churches, all the critical groups stand together in resistance. We do not think that this resistance will be fruitless. The radicalism we need in Canada, whether religious or secular, is one that promotes solidarity among all those groups who are disadvantaged in society and all Canadians who love justice, justice in Canada and justice for the less developed world.

### *CHURCHES MUST BE INVOLVED*

My theory-and-practice criticism of Berry's cultural analysis also applies to his religious position. It seems to me that churches become agents of political and cultural change only if their prophets, their daring thinkers, their innovators, speak from the center of the tradition. Their re-interpretation must verify itself in the religious experience of the people. The people must recognize in the new position the religion they have inherited. While I fully recognize the need for dialogue, brotherhood (sisterhood), and cooperation between Christians and members of other religions, I am not prepared to give up the Christo-centric perspective we have inherited. After all, we are just beginning the process of ecumenical association.

In my own writings concerned with the need to make theological room for Jewish religion, especially after the sad history of Christian contempt for Jews and the church's silence during the genocide inflicted on Jews by the Nazis, I have always proposed radical theories in a way that one could imagine them being integrated into liturgy and expressed in catechisms. Without this effort, I fear, religious radicalism does not make a difference in public life. Radicalism may be important for an individual, and I have great sympathy for personal freedom, but radicalism makes a contribution to social transformation only when it becomes institutionalized or finds a home in a tradition.

I have enjoyed Thomas Berry's essays. They have challenged my own critical perspective. I am grateful to the author for proposing his ideas to a wider audience. In a society characterized by apathy and indifference, possibly a veneer for fear and anxiety, radical ideas are precious. Still, we need radical ideas that feed collective action. I have not given up hope for a political movement of solidarity—of the kind described in the social teaching of our own bishops.

## *QUESTIONS FOR DISCUSSION AND REFLECTION*

1. Baum obviously considers Berry's ideas interesting, but not rooted in political reality. Do you find this concern plausible? Why or why not?

2. Do you recognize any strong elements of the Christian tradition in Berry's thought, or is he speaking only out of a secular, new-age perspective?

3. Baum claims that Berry is repudiating the Enlightenment. On what basis does he make this claim? Do you agree?

4. Do you believe that Berry has been radicalized by the American political situation and the American empire? Explain.

5. Both Berry and Baum find the bio-regional concept promising. How does this concept relate to a socialist perspective? How does it relate to Berry's "planetary socialism"? (See also Chapter Two.)

# 5

# *PATRIARCHY:*

## *The Root of Alienation from the Earth?*

Margaret Brennan

*Margaret Brennan, I.H.M., is professor of theology at Regis College, Toronto School of Theology. While she is sympathetic with Berry's outlook, she is concerned that he has not made a direct connection between his concerns and feminist issues. Her paper suggests ways of amplifying his analysis to include a treatment of patriarchy, of the relation between feminism and earth/matter symbolism, and the feminization of poverty. "The starting point of our discussion of the earth deficit," Brennan writes, "must reflect the historical neglect of the feminine principle while magnifying masculine values." Women's experience empowers them to discern the systemic and psychic links between various forms of injustice and to reverse the "mythic drive" to "continuing economic growth."*

For some years now, I have been challenged by the thought of Thomas Berry. His creative imagination always opens new possibilities and raises new questions. In the past I have found that fresh insights arise out of what is uppermost in my mind. Previous articles by Berry have challenged my views of spirituality, ministry, and the future direction of religious life, even when these areas were not specifically referred to. The two articles, which I have read in connection with this project, "The Earth: A New Context for Religious Unity," and "Economics: Its Effects on the Life Systems of the World," speak to me from a feminist perspective.

I find that Berry, particularly in the article, "Economics: Its Effects on the Life Systems of the World," writes from a position that touches a feminist issue in a very direct way. I am both intrigued and surprised that he has not made this connection. The constraints of time have not allowed me to develop all the aspects of this omission. Consequently, the following delineation of points is partial and incomplete, but may add to the discussion. I make them under three headings: 1) Patriarchy, 2) the Symbol of the Feminine as Related to Earth/Matter, 3) An Exhausted Economy and the Feminization of Poverty.

### *PATRIARCHY*

Whereas Berry posits the origin of the problem of economics in the industrial age, our current deficit in nature is as old as recorded history—which reflects a patriarchal worldview. History reveals a neglect of the feminine principle while magnifying masculine values (Gerda Lerner, *The Creation of Patriarchy* [London: Oxford University Press, 1986]). The starting point of our discussion of the earth deficit, therefore, must be moved to reflect this reality. Feminist scholar Sheila Collins, in an article entitled "Their personal is Political," points out that racism, sexism, class exploitation, and ecological destruction are four interlocking pillars upon which the structure of patriarchy rests.

The feminist experience has enabled women to penetrate the superficial differences to see the systemic and psychic links between the various forms of injustice. Exploring the patriarchal worldview, one can find metaphors that help to explain and connect the various manifestations of social sin. Such metaphors are summed up in a series of dualisms, the two halves of which are related to each other as superior to inferior, superordinate to subordinate, male/female, mind/body, subject/objects, man/nature, inner/outer, rational/irrational. These are the prime metaphors by which patriarchy orders reality.

The dismantling of patriarchal systems of domination and exploitation through the recovery of the feminine in human consciousness and history is the starting point for an authentic critique of contemporary social structures. The subjugation of the female by the male is the primary psychic model for all parallel oppressions between social classes, races, and nations.

Berry pleads for an end to the human exploitation of nature. If it is to be effective, this plea must urge the eradication of patriarchy which authenticates various human oppressions as well as those economic and political policies that encourage the rape of the earth. To call for anything less is to beg for a partial solution at best.

### *THE SYMBOL OF THE FEMININE AS RELATED TO EARTH/MATTER*

Berry has articulated an important insight about economics and the world needing a cosmological context for reinterpretation. Yet I believe his use of the term "human" refers essentially to a male perspective. The male paradigm is offered as the norm; the female perspective is conspicuously silent. Women, like the earth, have been a symbol of that which can be objectified, dominated, controlled, raped, plundered, and devalued (cf. Susan Griffin, *Woman and Nature: The Roaring Inside Her* [New York: Harper & Row, 1979]). To neglect this relationship historically and to fail to attend to the call to renew and reawaken the image of the feminine as it is articulated by contemporary Christian feminism is to ignore a new level of consciousness.

Listed among a number of authors Berry cites, indicating that the earth cannot long endure present modes of human exploitation, is Carolyn Merchant, who wrote *The Death of Nature*. Berry seems to have attended little to the first line of her introduction: "Woman and nature have an age old association" (p. xv). The whole purpose of Merchant's book is to point out the rooted connection between the women's movement and the ecological movement. Merchant states explicitly that her purpose is *not* to reinstate nature as the mother of humankind nor to advocate that women reassume the role of nurturer dictated by that historical identity. Her intent is instead to examine the values associated with the images of women and nature as they relate to the formation of our modern world and their implications for our lives today.

In investigating the roots of our current environmental dilemma and its connections to science, technology, and economics, we must re-examine the formation of a scientific world view that, by reconceptualizing reality as a *machine* rather than a living organism, sanctioned the domination of both nature and women (Merchant, p. xvii). It surprises me that Berry has not alluded to these assertions. Berry's use of the terms "human" and "nature" might be reworked in order to embrace the feminine principle. To omit this is to settle for half of reality which leads to an imbalance whose current manifestations are amply set forth in Berry's description of the earth's ills.

### *AN EXHAUSTED ECONOMY AND THE FEMINIZATION OF POVERTY*

Berry explicates at length on our crisis: an exhausted planet means an exhausted economy, both held captive to the mythic drive of continuing growth. But he does not refer to the stratification of the economy which has resulted in the feminization of poverty on a world scale. Would a functional cosmology such as he prescribes cure the ills of male domination? Or of class domination? Only a strategy with a dual focus will reverse the dynamic of the mythic drive. We must aim to change the nature of class, economic relationships, and relationships between men and women, as well as our attitude to the earth.

What we experience at every level: economic, social, political, is the result of a world ruled by the patriarchal mode. This is a dynamic of domination in which profits and property have rights over people. Women especially have suffered by having their influence defined and limited by men. For example, over the centuries, women have had experience in 1) sharing resources equably among family members; 2) taking the needs of individuals as well as the common good into consideration; 3) resolving conflicts; and 4) looking after relations with relatives and neighbors. But women's skills in these areas have been more and more confined because the microcosm of the family as a social unit has been dispersed. Many of its real functions have been transferred to business, industry, and public institutions. Women in the last four centuries have lost most of their real opportunities to influence events through their own work and abilities (cf. Ivan Illich, *Gender* [New York: Pantheon, 1982, 1985]).

Another way of looking at this situation is to recognize that the present non-monetary economy is largely a female economy. The real value of the informal economy in households, which includes the growing of food and its processing, education, running a household, etc. needs to be reclaimed. When these functions are performed within families they cost time and work. When produced outside families they cost money. This non-monetary economy is invisible. As such, it is an important manifestation of women's invisibility in society at large. In addition, such immaterial human needs as mutual respect, dignity, meaningful work and life, tenderness, caring, nurturance, human relations, have been given second place to economic growth and development. Satisfaction of these needs has been devalued and relegated to women.

### *THE MOST IMPORTANT FOCUS*

Feminism is perhaps the most important focus for changing the dynamics of domination because it transforms the way human persons see and relate to each other and the world. A feminist proposal would have us work towards an economy that not only recognizes and includes the earth's natural economy but also the non-monetary human economy as well. We need to link economy closely to the realities of human life and society. We need to restructure it in such a way that it serves real human needs (not competitiveness and control) and social interests. We need to bring the essential functions of life together: to integrate work, nurturance, and home. Concrete ways of beginning this process include: 1) all men and women working in both the private and public sector; 2) women and men working under the same terms in both sectors; 3) women having their fair share of participation in public affairs on their own terms, not through adaptation to male terms.

Finally I would like to address some remarks to Thomas Berry's paper, "A New Context." Berry challenges religious people, particularly Christians, to revise the imperialistic way they have defined their universality. By not absolutizing one particularism, Christianity, he calls us to embrace existing human pluralism. He spells out the consequences and foolishness of trying to fit all peoples into the mold of religion and culture generated from one historical experience.

A similar position is held by many Christian feminists. For example, Rosemary Ruether in *Disputed Questions: On Being a Christian* (Nashville: Abingdon, 1982), has said: "My preference for biblical thought is a relative preference for certain lines of religious vision that are characteristically Hebraic, not an absolute preference that rules out true knowledge of God/ess in other places. Nor do I accept the common liberal Christian distinction between particular or historical revelation (higher) and general or natural revelation (lower). All religions are rooted in particular encounters with God/ess and so are in that sense historical, although they may not necessarily make the historical itself a datum of religious experience" (p. 31).

With regard to pluralism, Ruether says,

> There is no final perspective on salvation available through the identity of only one people, although each people's revelatory point of reference expresses the universal in different contexts. Just as each human language points more or less adequately to universal truths, and yet is itself the product of very particular peoples and their histories, so religions are equally bearers of universal truth, and yet are particular in form. To impose one religion on every person flattens and impoverishes the wealth of human interaction with God, such as imposing one language on everyone steals other peoples' culture and memories (p. 67).

I would like to reflect with Berry on what can be done about a theology that not only assumes male standards of normative humanity but also is filled with an ideological bias that assumes women *qua women* are secondary and inferior members of the human species. Also, a major question in the whole area of religion is the tendency to make the divine a confirming theophany of the existing social order. The history of Catholicism has shown these pernicious tendencies at various times. Can they really be detached from vital, life-giving revelation?

*QUESTIONS FOR DISCUSSION AND REFLECTION*

1. Brennan identifies patriarchy as fundamental to the present global crisis. Berry does not use the term in his first two chapters in this book but he does in the final chapter. Are the ideas of Brennan and Berry compatible? In what ways? In what ways do they differ?

2. In your opinion, do feminists have grounds for relating the subjugation of nature to the subjugation of women?

3. What are some common elements in Thomas Berry's thought and that of contemporary feminists?

4. In view of our pressing global predicament, how can a feminist critique and Berry's critique mutually enhance each other?

5. At the end of her paper, Brennan writes: "A major question in the whole area of religion is the tendency to make the divine a confirming theophany of the exisiting social order. The history of Catholicism has shown these pernicious tendencies at various times. Can they really be detached from vital, life-giving revelation?" What is your reaction to this question?

6. Brennan also quotes Rosemary Ruether at the end of her paper. How do the ideas of Ruether and Berry compare?

# 6

# *REDEMPTION:*

## *Fundamental to the Story*

James Farris

*James Farris teaches philosophy of religion at Knox College and the Toronto School of Theology. His background in philosophical theology and his interest in science and theology are evident in the breadth of scholarship he displays in this essay. He is the only contributor who is not Roman Catholic, and his sympathy/distance from that tradition is helpful to understanding the whole process of Thomas Berry's cosmology. Farris' essay is both an elaboration of Berry's critique of modern alienation from nature and a dissent about what Farris regards as Berry's excessive optimism. The preoccupation with redemption, which Berry criticizes for distracting people from creation, remains for Farris a fundamental theme.*

My reactions to Thomas Berry's papers are those of a Protestant reflecting a philosophical orientation in theology. My most general response is one of sincere appreciation for the mind-stretching and imaginative insights that characterize Berry's writing, so far as I have encountered them.

I can affirm the validity of his call to move from objectifying modes of thought that yield only perspectival *Weltanschauungen*, toward a perception of the cosmos as the primordial subject which yields us self-understanding and our vocation as participants in and agents of cosmic process. I find his exposé and analysis of the degradation of our planet by the aggressive and exploitative projects of our industrial culture and its technologies to be disturbingly convincing.

He makes our economic systems themselves to appear as the most pervasive and invasive technological assault upon the sustainability of the global ecosphere. Our modern industrial society has regarded nature as plastic to our wishes, reshaping it in terms of technological structures of power. The power is asserted exponentially in quantitative terms rather than coherently in support of human symbiosis with nature.

I agree with Thomas Berry that in both religious and secular idealisms we have invested our mastery over nature with the attributes of spiritual fulfillment, celebrating the primacy of spirit over the opaque and alien forces of nature. From such expressions as "the Protestant work ethic" to "the conquering of space" we may reify this presumption.

Alongside the fragmented and partitive character of our industrial and technological insertions into the natural order, I would point also to the loss of holistic vision in recent philosophical and cultural criticism. Both linguistic and phenomenological analysis in recent European and North American philosophy have considered it liberating to be freed from the metaphysical system or the constraints of natural law. In these approaches to cognition, the appropriateness of language-games to given forms of life or the authenticity of diverse modes of being may be described or analyzed, but never called to account because there is no perceived paradigm of accountability.

A prevailing feature of modern consciousness has been its binding to textuality. Here I am referring to the text as the fundamental mode of signification. And I am distinguishing it from mythopoeic recital and rhetorical utterance, which are relatively primordial by contrast. These are varieties of texts: aesthetic, scientific description, philosophical or juridical propositions, all of which we read from our own text, the text of our self-understanding which has accrued through manifold readings.

What we too often fail to recognize is that the "world" is *pre-textual.* It is the source of our texts. Our hermeneutical endeavours have too often assumed that the interpretation lies hidden in the text, to be wrested from it, rather than to arise in a dialectical relation between the world of the text and the broadest horizon of our self-understanding, in our "worldliness." In the case of the religious text I would express this emergence of meaning as being evoked by a reading of the *textus receptus* in relation to the continuing experience of transcendence.

### *MESMERIZED BY OUR TEXTS*

Our failing is that we tend to become mesmerized by our texts; not least has this been true of Protestant fundamentalism's regard for the Bible. The classical authority that we tend to assign to written texts militates against re-readings and re-interpretations that take account of the paradigm shifts that are dictated by historical transitions and the fruit of intellectual enquiry into natural process. Moreover, in Christian terms, there has been a failure to pursue the complementary implications of the symbols: Word and Spirit.

It is helpful to be reminded by Paul Ricoeur that actions and events may properly be regarded as texts, because in a manner of speaking we may refer to them as "making their mark on time," tracing patterns of meaning. On this view, we might say that it has been our lack of attention to the hermeneutics of action that has obscured for us the implications of the move to a global economy and the growing evidence that the fate of humanity is bound up with the fate of the planet. We remain tied to scripts that echo conventional wisdom, such as the alleged necessity of fostering continued economic growth.

Not all of the trends in contemporary and literary criticism remain stuck in preoccupation with the delimited and constricted phase of human understanding, i.e., the microphase. There are some significant cognitive moves being made beyond mere apprehension, toward comprehension. In addition to organic and holistic philosophies associated with such figures as A.N. Whitehead or M. Polanyi, we may name a recent convert from the ranks of analytical philosophers of science in the person of Stephen Toulmin.

In *The Return to Cosmology: Postmodern Science and the Theology of Nature* (Berkeley: University of California Press, 1982), Toulmin hails the debunking of scientific myths, including the assumption that the scientist is properly and exclusively an observer. And he assesses appreciatively the trend to cosmology in the thought of Koestler, Teilhard, Monod, Sagan, and others. He welcomes the emergence of a green philosophy which supports a participatory, ecologically responsible ethic, and offers encouragement to a renewed theology of nature.

### *MUCH LESS SANGUINE*

In the foregoing, I believe I may have demonstrated my general resonance with Thomas Berry's analysis. I am much less sanguine than he appears to be that a paradigm shift, replacing a masterful domineering attitude toward the earth with one which accepts the originating, nurturing environment of the earth, will enable us to write a new human story, engendering the macrophase of our moral and spiritual potential. On the basis of steps we have heretofore taken on the road toward comprehensive global or cosmological consciousness, I am not inspired with confidence that now we are ready, by taking thought, to achieve a bonding of our vocations and our institutions with "the earth as the primary self-propagating, self-nourishing, self-educating, self-governing, self-healing and self-fulfilling community."

My reservations stem largely from the claim advanced that we have been hindered on our path toward such beatitude in Christian and western culture by the theological preoccupation with redemption at the expense of creation. I grant that this is justifiable criticism of a major strand in Christian dogma and piety where deliverance from the constraints of the flesh and of the present evil world has been the prevailing concept of redemption. This owes more to the influence of Greek rather than Hebraic categories, and, in millennarian fundamentalism, to an impoverished, literalistic view of the Bible.

I believe that "redemption" remains a fundamental theme in religion. The primary term of religious experience is not coherence but transcendence. In religious terms, coherence comes by way of a fulfillment that is reconciliation with the perceived transcendence. "Reconciliation" is not only the favoured New Testament, pauline symbol for redemption, perceived in both personal and cosmic terms, but it is also expressive of the quest for cosmic consciousness in eastern religions.

### *A DIMENSIONAL DIFFERENCE*

Experiencing the transcendent and relating to the transcendent do not appear to me to be the same thing as entering the macrophase of consciousness, the latter term employing a spatial metaphor. It is not simply a case of being "magnanimous" in the etymological sense of that term. It is rather a sense of dimensional difference in which we become aware of our finitude, and of our finitude as being somehow *de-fined* for us. As the self-conscious children of cosmic dust, we are made painfully aware throughout our history of our moral ambiguity. As the offspring of natural process, we are nevertheless not spared from natural-or-moral-catastrophies by reverencing nature. I hardly need to identify by name one modern European society which perfected the technology of death while evoking the elemental symbols of blood and soil to inspire its millennial vision.

If our spirituality, and with it our morality, is to be informed by cosmic vision, we must at least acknowledge that the path toward this vision has entailed revision, discrimination, and self-judgment on our part. Cosmologies fashioned in the past history of science, philosophy, and theology have not notably spared their architects and attendant societies from cosmic discomforts or moral failures.

I would suggest that here is an imminent danger of succumbing to the "naturalistic fallacy" in assuming that consciousness-raising about natural states of affairs will assure that we become compliant with nature's nurture and its self-regulating ends. If the destiny is assured, it hardly needs pointing out; if it is problematic or ambiguous, then we must look to some other ground of hope, or accept the absurdity of our fate.

It is at this point that I wish finally to comment briefly on the discussion of the role of world religions in the human community when socialized within a planetary economy. Berry's vision of the role of existing religious traditions is a modest one, and admittedly there are strong historical grounds for this reticence. The gift of complementary religious traditions to planetary consciousness is seen by Berry to be less promising than the fruits of the scientific meditations upon nature.

I find anomalous the view that the religions must tarry meanwhile at the microphase, in light of the fact that the genius of religion has traditionally been seen in the power to relate individuals and societies to ultimacy and completion, however symbolized. If the complementarity of religions holds promise of eventually informing a more adequate spirituality of the natural order, surely it must involve discrimination among the traditions and the visions they reflect.

The scientific meditation has perceived the world in terms of a discriminating cognitivity and has thereby achieved disclosure and enlarged understanding. It is an undue constriction to lay upon religion to assert that the testimonies of all religions, differing as they do in cultural, cognitive, and cosmic outlook, should be valued indiscriminately so far as they fit the new story authored under the scientific muse. My own conviction remains that the Judaeo-Christian tradition offers the most promising motifs and the broadest vision to illumine the macrophase of planetary existence, and its beyond.

## *QUESTIONS FOR DISCUSSION AND REFLECTION*

1. In your opinion, are humans capable of greater evil than other species of the universe? How would you justify your stance?

2. Farris raises the question of redemption and its role in the Christian understanding of life. Do you think the natural world needs redemption? What does this mean?

3. Do you think there is a danger that Thomas Berry is leaning towards an "illumination" model of what is required for the earth, rather than a "conversion" model? In what does this danger lie?

4. Farris writes, ". . . I am not inspired with confidence that now we are ready, by taking thought, to achieve a bonding of our vocations and our institutions with 'the earth as the primary self-propagating, self-nourishing, self-educating, self-governing, self-healing, and self-fulfilling community." What is your reaction to this statement?

5. According to Farris, "My own conviction remains that the Judaeo-Christian tradition offers the most promising motifs and the broadest vision to illumine the macrophase of planetary existence, and its beyond." Do you agree or disagree? Why?

# 7

# *NEEDED:*

## *A New Genre for Moral Theology*

Stephen G. Dunn, C.P.

*Of all the theologians represented in this volume, Stephen Dunn is the one who has most systematically explored the implications of Berry's thought for a specific theological discipline. A moral theologian, Dunn has built many of his courses at St. Michael's, Toronto School of Theology, around the premise that the earth is a subject for theology. For Dunn, creation is a teleological horizon, and the religious dimension of life must be thought of as advocating participation in earth's time-developmental processes. What one has a sense of in this paper is the excitement of a new vision informing moral theology that ultimately, Dunn hopes, will help wisdom develop in many disputed areas, such as human sexuality, biotechnology, and structural and personal evil.*

The Book of Ecclesiastes says there is nothing new under the sun. Yet, from time to time, the human spirit does actually catch a glimpse of something genuinely novel. Thomas Berry has sighted something in the stars, something novel, that makes him affirm humans as "articulated stardust." He has seen cosmic time, and he has pushed ahead to formulate a functional cosmology that can serve both scientists and theologians. Knowing something new, for

Thomas Berry, has stimulated a new dimension of knowing.

This new dimension is the concept of macrophase and microphase reality. Through Berry's work, the old wisdom is retrieved within a new critical context, which will help theology out of its hidden corners into the open where it can speak to our culture. Before discussing Berry's new dimension any further, however, let's first take a look at moral theology, which has been groping around for the last twenty years in a changed world.

## *A BRIEF HISTORY*

Early post-conciliar enthusiasm for Christology, as a new and invigorating matrix for the renewal of moral theology, soon gave way to the burdens of modernity. With the demise of the manuals, efforts to relocate the moral enterprise within received religious language showed much promise (cf. Gérard Gilleman, *The Primacy of Charity* [Westminster, Md.: Newman Press, 1959]), but also could not provide the breadth needed in a church committed to participation in the world as it was outlined (sometimes naively) in *Gaudium et Spes*.

The significant post-counciliar venture for moral theology was to take moral philosophy much more seriously. With Reinhold Neibuhr, we began to speak of an ethics of responsibility. We espoused the cause of personalism to escape a heritage of legalism. Catholic moralists began to worry over deontological implications of traditional moral thought. Teleology seemed more congenial, but factually ranked such moralists with pragmatists.

Leaders like Bernard Häring, who had confronted the Kantian question of autonomy and taken the world of sociology seriously, attempted to put new wine in old wineskins of traditional vocabulary. That, too, missed the mark as a new Catholic coherence. With Jürgen Moltmann we learned that western culture had privatized religion and that the believing, hoping community ought to be building the City of God with the Cities of the Earth. We became cognizant of Marxist perspectives. With Jacques Ellul, we were nudged toward a more critical sense of the perils of technology. Where has all this led the discipline of Catholic moral theology?

With Bernard Lonergan, we see we have left the classicist world and truly espoused historicist thinking. We have achieved the turn to the subject. But Lonergan's sardonic remark about arriving breathlessly at a point the rest of the world is leaving, is already proving prophetic. Matthew Lamb would try to move Lonerganian categories into contemporary significance through insisting on the "subject in history" (*Solidarity with Victims* [New York: Crossroad, 1982]). Latin American theologians would move the moral subject further into liberation praxis. North American liberation theologians, too, would extend the moral theological enterprise, especially with black and feminist theologies and the concern for the Third World.

Meanwhile, this ever-accelerating momentum toward the expansion of the Catholic conscience and imagination has brought forward a new emphasis on biblical image over intellectual discourse in the hierarchy (post-conciliar encyclicals, Canadian and American national bishops' conferences, and papal statements), the most forceful of which currently is "the preferential option for the poor."

Where has this left the discipline of Catholic moral theology? In my view, mostly out in the cold. Moral theology's intellectual categories are not yet at the point of consensus. Some religious language tends to confuse rather than clarify, since the tradition of Catholic moral theology has been an intellectual one. The Catholic mind wants to be explicit about the Marxist thought in *Laborem Exercens*, or the ideology of progress tucked into some paragraphs of *Gaudium et Spes*, or some pages of Teilhard. More than anything, I think, the Catholic tradition desires an intellectual matrix that is comprehensive (a legacy of the now out-dated natural law tradition), a matrix that is religiously convincing (a legacy of outmoded deontology), a matrix of cultural inclusion (a legacy of medieval syntheses).

### *A FUNCTIONAL COSMOLOGY*

With that as the current agenda, it is exciting to probe Thomas Berry's articulation of a functional cosmology. For the promise is that within a matrix that is comprehensive, religiously convincing, and culturally inclusive, we may perhaps see moral theology address the fundamental question: "What does it mean to be human on this planet?"

I will tentatively suggest here where Berry's insight into the macrophase and microphase nature of the universe will be helpful. First, Kantian subjectivity and autonomy are primarily valid in macrophase: the Earth as Subject (cf. the interesting developments around James Lovelock's Gaia hypothesis (*Gaia: A New Look at Life on Earth* [London: Oxford University Press, 1979]), or *The Anthropic Cosmological Principle* by John Barrows and Frank Tipler [Oxford: Clarendon Press, 1986]).

The famous phrase of Dobzhansky, that the universe is not random, nor purposeful in the Aristotelian sense, "but groping," takes us into a new sense of design, not mechanical, but closer to artistic processes. From this macrophase perspective, the human is "that being in whom the universe reflects on itself" in Berry's phrase, "and celebrates itself in conscious self-awareness." This, of course, leads one to a much different perspective on microphase issues: We are first of all a species of the earth and our responsibilities and challenges are far greater than we have traditionally reflected upon (cf. Mary Midgley, *Evolution as a Religion* [New York: Methuen Inc., 1986] and Hans Jonas, *The Imperative of Responsibility* [Chicago: University of Chicago Press, 1985]).

Secondly, the religious perspective is deeply enhanced on the macrophase level: Creation is the primary revelation of the divine, and all human cultures have something to say to the revelatory nature of this universe. The autonomous ethic, already challenged by liberation theology, subsists in microphase realities of culture, community, social structures, and personal levels. The creation of culture and individual lives are not determined by necessity, but exist in responsible personal decisions. But the depth of our interconnection with the whole earth is a much more primary reality than our autonomy.

In macrophase, our traditional western preoccupation with history broadens out to a time-developmental mode of consciousness. Instead of basing our microphase ethics on a static universe, we have primary questions: What is the universe doing? What is the earth doing? Reflections such as Berry's "Twelve Principles" teach us in a way we need today, far more than wisdom culled from other ages and times that did not face the questions we are facing.

In macrophase again, the realization that the cosmos is intelligent leads us to see in microphase that science, as "the yoga of the west" in Berry's phrase, can be self-critically a guide to ethical inquiry. And finally, as we awaken to a cosmos that is our universal macrophase, in microphase a particularly opaque evolutionary process justifies critical human pragmatism, also "groping" as the cosmos does. But retrieval of lost intellectual and religious energies is not the whole story. New energy is also generated by the functional cosmology Berry introduces to western thought.

### *A UNITY OF AFFIRMATION*

By describing the cosmos in macrophase/microphase, we retain a unity of affirmation of reality we have not before experienced in moral theology. It means that we can do for ethics what Berry insists Christianity did for the concept of God. Trinity, in this context, is a macrophase concept. Once its reality is asserted, divinity could no longer be absolutized outside of relation.

On a different level, we commonly affirm macrophase reality when we speak of community and person-in-community; something larger than person (though not destructive of it) is announced in the word "community." Something defining the person (though not univocally) is announced in the word "person" when community is the context. Berry is saying that macrophase is always the primary context. This is a blow to the style of the antonomous ethics we have known in the modern period, but it is a liberating expansion of the sense of the real that is sorely needed. It is a paradigm shift to be sure, bringing with it implications that I will now summarize as a conclusion:

1) The community of all earth creatures is primary to ethical thought.

2) God's creation must be thought of as on-going.

3) Creation as a theological horizon must be more in evidence in ethical deliberation.

4) The religious dimension of life must be thought of as advocating participation in earth's time developmental processes. Natural law's "participation in the mind of God" becomes participation in the work of creation, a mysticism of earth participation, grounded in a deep ecological religious ethic.

5) Christ in the mystery of the incarnation must be appreciated within the context of the macrophase reality of incarnation: participating in the evolution of the earth and the cosmos.

6) Christ in the mystery of redemption must be appreciated in terms of the totality of salvation history, which goes back in time to the beginning of the universe and forward to the evolutionary future.

7) Re-visioning, or what Berry calls establishing a "new genre," becomes an essential task for ethics. The interfacing of the macro- and microphases might yield us a context that will enable the human race to get on with the deeper solutions to the urgent problems of our time. Berry's work in economics gives us an example of the possibilities.

8) The earth's urgencies must be deemed our primary urgencies.

## *CONCLUSION*

My response to Thomas Berry then, from the viewpoint of moral theology, is to highlight his macrophase/microphase framework. He has accomplished what Vatican II aspired to achieve, transcending the dichotomous framework with which the Catholic community has been approaching the ethical and religious dimensions of reality. He therefore opens the door to an ethics that is participative, religious, and Christian.

His strength is in the macrophase; his challenge to disciplines such as moral theology is, having seen what he sees in the stars, to move forward with the pressing issues—perhaps first and foremost by developing the "new genres," trusting the power of a truly functional cosmology to be morally persuasive.

## *QUESTIONS FOR DISCUSSION AND REFLECTION*

1. If you were to teach a course in ethics to college students who had never had ethics before, which of these topics would you address as being fundamental: the use of nuclear power; authority and obedience; family life; the natural law tradition; the consumer society; population control; the dignity of the person; biotechnology; biocide (the elimination of other species); sexuality; option for the poor; Third World oppression; technology; the moral subject; geocide (the killing of the planet); other? Explain your choice or choices.

2. Many people who are concerned with social justice have a problem with ecology because they believe it is elitist and middle class (i.e., save the seals and forget the fishermen). Are these two perspectives mutually exclusive? How would you reconcile them?

3. Dunn states, "The depth of our inter-connection with the whole earth is a much more primary reality than our autonomy." Is there any issue in ethics that you can mention that would be helped by such a perspective?

4. Dunn believes that "macro" issues must be faced in ethics: "If we were to take the nuclear fact. . . it doesn't matter if you're Russian or American or Peruvian, you're dead, and what inspires the peace movement is facing the essential insight that nuclear war is mutually destructive. Beyond all other issues, the ecological situation is life and death. Even if we put all the nuclear weapons away, we are still killing ourselves." Do you agree or disagree that this issue should be dealt with in ethics courses? What are your reasons?

# 8

# *SCIENCE:*

## *A Partner in Creating the Vision*

Brian Swimme

*Brian Swimme, a physicist, has worked with Matthew Fox at the Institute in Culture and Creation Spirituality at Holy Names College, Oakland, California, for several years. In 1983 he took a one-year sabbatical to study with Thomas Berry, and has since collaborated with him in several projects. Currently they are writing a book together on Thomas Berry's "Twelve Principles." Swimme's most recent book, The Universe Is a Green Dragon, is dedicated to Berry.*

My own professional training is in mathematical physics, and in this chapter I would like to indicate what a scientist finds valuable and unique in Thomas Berry's work.

Before I get to Thomas Berry's contribution, however, I think it's important to bring to mind two facts about contemporary society. The first is the general apathy scientists feel toward religious thinking in general. I was reminded of this at a recent conference for Catholic scholars held at the University of California at Berkeley. One physicist came. And one mathematician. And this from a university that employs hundreds of scientists. The second point is that approximately half of the world's scientists and technologists are employed in war research and development. And how many of the remaining are employed by corporations heavily implicated in ecological destruction?

The context for my remarks is this: We are all suffering under the nuclear threat and ecological devastation, and yet our spiritual or ethical traditions seemingly stand by unable to make a difference. They do show up at the sidelines and moralize, but it always comes at the end of the game. This moral criticism is of great importance, surely, but it comes too late. It comes only after the science and technology have been taught. This moral stance is ambivalent at best; and it's true that the moralizing comes from the Catholic tradition, but Catholic universities prepare scientists and technologists for war research and ecological assault as well as anyone else.

What is missing, and what Thomas Berry provides, is a functional cosmology that will enable the human community to organize itself in a way aimed at planetary health. Nothing less than a comprehensive vision of the universe is required. A new social theory, a new psychological theory, a new economic program will make no impact on the scientific-technological trance that is behind our impasse.

Thomas Berry's achievement is to position himself within the knowledge that scientists and all the rest of us regard as obviously true. His starting point is natural selection and genetic mutation, the second law of thermodynamics, the initial singularity of spacetime, the innate releasing mechanisms of neurophysiological response. His starting point is the universe as it has been discovered by our contemporary scientific modes of understanding. By taking the universe as primary, he is able to work out a cosmology that is meaningful to anyone educated in modern ways of knowing. It is this cosmological achievement that must be understood if the full significance of Thomas Berry's work is to be appreciated.

Before going further into a discussion of the main outlines of his cosmology, I think it might be helpful to say briefly what Thomas Berry does not do. I do not mean to imply that these other approaches are without value. I am simply trying to clarify what is unique in Thomas Berry's thought.

1) He does not set out to prove that religion and science are compatible. There are a number of thinkers involved with this enter-

prise. They examine the methodologies and epistemological assumptions of scientists and theologians, and arrive at subtle distinctions concerning truth claims and so forth. All of this is absent from Thomas Berry's work.

2) He is not interested in adjusting the world of the sacred to fit scientific categories of thought. Such a program has been carried on throughout the scientific period, based on the assumption that science is the whole truth and needs to eliminate the superstitious claims of religion and so forth, all ending in an unacceptable belittling of reality.

3) He does not translate the universe into theological or scriptural modes of thought. This enterprise, too, is common, and one that irritates scientists as much as number two above irritates religious people.

### *WHAT DOES BERRY DO?*

Thomas Berry wonders over the revelations of the universe. He wonders over the human. He faces the great discoveries of tectonic movements, of spacetime curvature, of stellar nucleosynthesis and he asks himself: What does all this reveal about the role of the human species in the universe?

Here's the great surprise: *Thomas Berry teaches scientists about the universe.* I would say that this is why a scientist encountering his work for the first time grows ecstatic, as if in the rush of an illuminating event. Thomas Berry provides not psychological insight, nor moralistic guidance, not mystical or contemplative teaching about withdrawal and detachment from the world, but knowledge and insight into the nature of the universe. This is such a shock, such a disorienting surprise, a scientist can hardly believe it is actually occurring. One has become so accustomed to assuming the universe must always be left out of religious thinking.

I think more should be said here, or misunderstanding will be inevitable. It is true that most or all theologians speak of "reality" or the "world" or "life." But scientists have an automatic question: Do theologians have a great deal to say about life if they are unaware of the dynamics of cell division, of genetic language, of the elegance of photosynthesis? Should a person use the word "life" if he or she knows little or nothing of the actual events comprising the four billion year epic of life on this planet? From a scientist's point of view the theological use of the word "life" appears to be something spiritual or psychological, something perhaps germane to moral statements, but disconnected from the structures of the actual world as it is given to us in our most basic investigations.

Here is Thomas Berry's point of view: Theologians, when they speak of "life" in a moral, spiritual, or psychological sense only, inevitably scale down their own theology. Thomas Berry criticizes the theological enterprise today for its timidity. The scriptural "I am the way and the life" has been crippled to mean little or nothing at all, precisely because theologians, when they say "life," do not mean "biological life." It is to overcome this parochial attitude that Thomas Berry offers a comprehensive interpretation of the universe, one that goes beyond both science and theology in and of themselves.

I have said that scientists learn about the universe from Thomas Berry, and this accounts for their enthusiasm for his thinking. But that is not all. There is something else that is vital. Scientists recognize Thomas Berry as a brother—as one who has discovered and been stunned by the beauty of the universe. He is just as amazed as they are by the universe. This alone enables a scientist to value Thomas Berry's intuitions. For here is a person—even a religious personality—who is as devoted as they are to the beauty that suffuses the world.

I think it is important to say something about how rare this is, and how essential. It explains both Thomas Berry's accomplishment, and the inadequacy of so much theology of recent centuries. Scientists have learned that it is not wise to speak to religious people about one's devotion to the universe or nature. What happens generally is that theological, clerical, or spiritual people, without thinking at all, belittle such devotion. How do they do this? By transforming the great majesty of what a scientist is trying to express into side comments on this or that scriptural text, this or that theological doctrine. They listen to this excitement unaware that they listen within a theological orientation, and this inevitably leads them to discount it as secondary.

Why do scientists have little interest in theologians? When you fall in love you bring your beloved home, beaming with excitement; but if your parents disapprove, if they are disappointed, if they are apathetic or if they engage in psychological assault, you eventually learn to stop dropping by the old homestead. Scientists dropped out of religion because theologians and preachers had nothing interesting to say about the universe. Then, too, they stopped sitting in the pews because the preachers kept explaining to them that their passions and interests, their meaning, their central devotions in life are unimportant—or irrelevant and footnotes to the real truth.

For Thomas Berry the universe is primary. He enters with no distracting agendas drawn from conciliar documents. He does not attempt to see the universe as a gloss on the Bible. From his point of view, to attempt to cram this stupendous universe into categories of thought fit for scriptural studies or systematic theology is to lose the very magnificence that stuns us in the first place. Our encounter with the universe must be primary, for the universe is primary. In his view, the stars, the mountain ranges, and the clusters of galaxies demand and are worthy of our deepest regard. Our attention must be turned to the vast drama and majesty of the universe if we are to discover our role at the species level of life.

I would like to describe here some of the main features of Thomas Berry's cosmology. At best I can give only a dim picture of the whole vision he offers.

*1) The great achievement of the scientific era is the cosmic creation story*. It's interesting to me that in my graduate courses in mathematics and physics, in my conversation with scientists, and in my readings in the philosophy of science, I never once learned the full significance of what we scientists were doing. Embroiled in the work, most scientists did not have the freedom to recognize the cosmic story as their aim. There were some thinkers who did, of course. Albert Einstein explained that science was essentially "the recapitulation, in the conceptual realm, of the universe."

But Thomas Berry goes beyond Einstein in emphasizing the cultural role this "recapitulation of the universe" will play. It required someone with a knowledge of the world's cultures back to the tribal period to recognize the transcultural and trans-scientific meaning of the story of the universe that burst into being some billions of years ago, that developed in complexity and form throughout the galactic, planetary, life, and human stages. Though scientific knowledge has put lethal weapons in our hands, it has also provided the earth with the first common story of our origins and development. The scientific enterprise has eventuated in a creation myth that offers humanity a deeper realization of our bondedness, our profound communion not only within our species, but throughout the living and non-living universe.

*2) The new story is an empirically-based story.* Precisely because this story of the universe comes to us through our investigations beginning with our eyes and ears and body, can we speak of a transcultural creation story. Members of every continent are involved in discovering and articulating this story. Members of every major religious tradition are involved in its telling.

What other story could possibly have served humanity as a whole? How else could members of Hinduism and Christianity and Native American Traditions come to agree on the ultimate origin and development of the planet, of life forms, or of the stars, but through a direct experience of these realities? So long as humans insist on their own scriptural stories, they only emphasize their differences. But with a story that begins with the wind and sunlight and the continental movements, as they reveal themselves through our direct investigations, we have the promise of a convincing story, one around which we can work out a common reverence and a common set of values.

*3) We live in a time-developmental universe.* The larger meaning of time has been discovered in the last two centuries. It is difficult to estimate the full significance of this fact. The universe in which we live and think is vastly different form the spatial or cyclical universes of Plato, Aristotle, Aquinas, or Shankara, for instance. For all previous humans, the universe was something set. Either all the species came into existence at the same time, or else they emerged as spring and summer emerge, in a cyclical pattern. In such worldviews, time is seen either as related to decay, or as unreal, a wheel of delusion.

In our new vision, time's dynamic reveals itself in an ongoing creativity. Only through the most prolonged meditation on the universe could this creative dimension of developmental time be appreciated. Eventually it was recognized that species were not set from the beginning, but were created throughout time. Eventually it was realized that the earth had not simply been here from the beginning, but was involved in a vast development stretching back billions of years. The awareness completed itself when physicists discovered that the universe as a whole was a self-emergent dynamic, a one-time energy event caught up in its own inner developments through time.

The insights of all thinkers previous to our time are, to varying degrees, conditioned by spatial cosmologies, all of which have been surpassed. Thomas Berry's insistence is that until we begin our thinking in this time-developmental universe, we condemn all our thoughts to conceptual frameworks in the midst of collapse. How convincing are theologies that are framed by worldviews no longer regarded as real? In insisting that we begin our thinking within a time-developmental universe, Thomas Berry continues an intellectual tradition best represented by Thomas Aquinas, when he set out to learn the philosophy of Aristotle. This tradition insists that our insights must be framed in our best knowledge of the universe. Aquinas had his Aristotle; Thomas Berry has his Newton, Darwin, Lyell, and Einstein.

*4) Everything in the universe is genetically related.* Humans and yeast are kin. They organize themselves chemically and biologically in nearly indistinguishable patterns of intelligent activity. They speak the same genetic language. And all things, whether living or not, are descendants of the supernova explosion. All that exists shapes the same energy that erupted into the universe at the primeval fireball.

No tribal myth, no matter how wild, ever imagined a more profound relationship connecting all things in an internal way right from the beginning of time. All thinking must begin with this cosmic genetic relatedness.

*5) The universe is integral with itself.* We have already mentioned the intrinsic connection between the human body and the star that created the elements of the human body five billion years ago. We should call to mind as well the cosmic background radiations that communicate events of fifteen billion years ago to humans who are alert to their intelligibility today. These are macrophase examples of what is most obvious to us today in our new ecological awareness: We live in an interconnected universe. Every being on earth is implicated in the functioning of the earth as a whole; and the earth as a whole is intrinsic to the functioning of any particular life system.

We have discovered that this integral nature of the universe extends back to the beginning of time. The conditions and dynamics of the fireball were such as to enable life to develop within the universe. We can even speak of the way in which the human face is there in the structures of the fireball, for if the elegance of the fireball were changed substantially, life and human presence would be eliminated. In this cosmological vision, life is more than a quality characterizing certain events on a particular planet. Life is a principle inherent to the primordial structures of the universe.

*6) Humanity is a celebratory species.* Rather than seeing human self-awareness as simply an addition to the planet, Thomas Berry describes a planet that becomes aware of itself through the human element. In this vision, the human emergence is an activation of a deep dimension of the universe. It is true of course that the human is an individual being on the planet; but it is equally true that the human person is a mode of the planetary process as a whole.

From this perspective, humans are activities of the earth. Scientists and artists are not simply exercising their private talents and rational capacities when they set about their work. Rather, in these very activities, the earth is simultaneously revealing and discovering itself. This earth must be understood as an ongoing developing activity that has eventuated in mountain ranges and humans and ants. The earthquake and the Mozart symphony are both activities of earth—not of Mars, not of Jupiter. They are processes grown out of the fundamental dynamism of this planet.

The universe as a whole is a great celebration of that ultimate mystery whence it came. But the human species is especially created for celebration, for in human awareness, the universe turns back on itself in admiration and joy.

*7) The three basic laws of the universe are differentiation, subjectivity, and communion.* The universe is differentiated. It comes to us in articulated energy constellations, not as a simple homogeneous material. The more closely we examine anything, the more clearly do we appreciate its unique differentiation from everything else that exists in the universe.

The universe consists of acting subjects. After our penetration into the deepest reaches of matter, we realized that there was no such thing as an inert thing; there was no matter that was not simultaneously churning with activity. An atom is a centered, self-organizing entity. The earth is a centered self-organizing entity—so too the virus, the galaxy, and the forest. The universe consists of subjects.

The universe is bound together in communion, each thing with all the rest. The gravitational bond unites all the galaxies; the electromagnetic interaction binds all the molecules; the genetic information connects all the generations of the ancestral tree of life. We live in interwoven layers of bondedness.

The planet as a whole is in a traumatized state. This is not the first time. Over four billion years, earth has repeatedly found itself in crisis moments, but each time a stupendous creativity has come forth enabling the planet to move from misery to health. The invention of photosynthesis might be the most spectacular. In my thinking, the creativity breaking through Thomas Berry represents that vision of the world that will enable the earth to move toward health.

## *QUESTIONS FOR DISCUSSION AND REFLECTION*

1. Brian Swimme believes that our spiritual and ethical traditions ignore the present nuclear threat and ecological devastation. What, if any, evidence have you found for awareness about current ecological devastation among religious spokespersons?

2. If, as Swimme claims, scientists are enchanted with the universe, and Christian theologians and leaders are not, what do you think causes this gulf? Is there any bridge across it?

3. "That all life on earth is genetically related" has caused con-

sternation since Darwin's time. Brian Swimme goes even further to claim that the whole earth has been shaped by the supernovas. Why do these claims cause so much difficulty in the various Christian denominations?

4. Swimme says that the scientific tradition has "provided the Earth with the first common story of our origins and development." Is this true or is this just another claim of western supremacy? What do you think?

5. Although Swimme states that we are at a crisis point in the history of the earth because of human devastation, he remains hopeful because the earth has moved through many life-and-death crises in its history. Are there ways in which you share this "earth-hope"? What are they?

6. Both Swimme and Berry find a role for humans in the celebration and conscious awareness of the universe. Recently, much debate has centered on "specie-ism": the arrogance of special claims for humans as against other species on the earth. In what ways would you argue for, or against, "specie-ism"?

# 9

# *THE NEW COSMOLOGY:*

## *What It Really Means*

Caroline Richards

*Caroline Richards holds a doctorate in history from Stanford University. She has been writer-in-residence and staff member at Holy Cross Centre for several years and presently teaches in the Peace and Global Studies Program at Earlham College in Indiana. She has written numerous articles on ethics, ecology, and current events in Latin America. She is co-editor of this book, with Anne Lonergan.*

Our problem, writes Thomas Berry, is that we have no community story, and no community can exist without a community story. Long ago, before the Black Death ravaged Europe, we had one, but that disaster, which killed half the population, permanently crippled the western psyche, causing some people to turn inward to an excessively private and personal form of spirituality and others to seek truth in a desacralized, objective search for scientific knowledge. While each approach provides certain satisfaction, neither is able to furnish adequate answers to our most basic questions about what humans are, their role in the world, and the meaning of existence.

Science empowers us with ever more sophisticated means to manipulate the natural world, and, possibly, to destroy it along with human life. By itself science gives us no wisdom. Religion, on the other hand, is a source of comfort for many, but its stories are sectarian stories, unconvincing to the non-believer, and they exist alongside and apart from the secular learning on which modern civilization is founded. Religion and science form separate worlds of discourse; neither, as presently conceived, is capable of giving us the wholistic vision vital to our survival.

### *THOMAS BERRY AND "THE NEW STORY"*

It is interesting that Thomas Berry wrote his study entitled "The New Story" in the 1970s, when, in the afterglow of Vatican II, scholars were remaking the Catholic theological landscape. Up to that time, the church had attempted to account for the world's existence and our duties toward it, toward each other, and to God, by reference to natural law theory based on the cosmology of Aristotle and later more fully developed by Thomas Aquinas. The theory could not, however, inspire the sort of life-giving vision that Thomas Berry would argue for without extensive reassessment in light of the modern predicament. By mid-century many Catholic theologians besides Berry, discouraged by what they had come to regard as a medieval and anachronistic system, concluded that the natural law theory was no longer persuasive. They believed that it was time to explore other ways of articulating the meaning of our faith. The many sophisticated, solid, and exciting works which have appeared in Catholic philosophy and theology since Vatican II attest to these explorations. In order to understand the unique place Thomas Berry holds in this intellectual landscape, it might be helpful to consider here, though briefly, some of the principal developments in recent Catholic thinking and their implications for the kind of cosmological vision Berry advocates.

### *THREE DOMINANT TRENDS*

Though there are obviously many others, we will focus here on three trends which have tended to dominate in Catholic intellectual circles in the past twenty years.

*1) A movement away from abstract thinking toward a ground-*

*ing of theology in experience.* Following the thought of Karl Barth, some Catholics claim that faith is not reasoned argument but rather a gift from God to which humans respond with obedience. The starting point of theology is God's self-disclosure in Scripture, in the person of Christ, and in current encounters with God's people. Yet others, influenced by existentialism, contrast knowing, experiencing subjects with the rest of creation. We are authentically human, they say, only when we are personally involved in concrete acts of judgment and decision-making—not when we are formulating abstract laws or systems.

The turn to the experience of the individual as the source of theological knowledge is perhaps the most readily apparent departure from traditional Thomistic theology among Catholic writers. It is found among Thomistic revisionists as well as among existentialists, neo-Kantians, Marxists, and others. For example, Karl Rahner (1904-1984), the doyen of Catholic theologians since Vatican II, took as his starting point the individual as person—creature of time and space, history and culture—who could, in his or her depths, encounter "Holy Mystery," the forgiving and loving God. In explaining why he was a Christian, Rahner began not with the traditional Thomistic proofs for God's existence, but with the question that is implicit in being an experiencing subject.

Many other Catholic thinkers, too, believe that the test no longer seems to be right belief so much as personal faith and trust—faith and trust which can, when appropriate, recognize the need for correction and reform in the tradition.

*2) A shift of focus from the redemption of the individual to the liberation of the community.* While European and Anglo-American theologians have effected a turn toward experience that is centered on the individual, Third World liberation theologians have enlarged the perspective to include the community. For them the primary Christian paradigm is not the individual standing lonely before God, but the Exodus community seeking liberation. As David Tracy has written, with the rise of liberation and political theologies, the theological landscape has been "irretrievably changed" ("Introduction" in *The Challenge of Liberation Theology: A First World Response* [Maryknoll: Orbis Books, 1981]). While earlier European and

American existentialist and transcendental theologies were prone to an unbiblical, apolitical individualism, liberation theologians have restored to Christianity the prophetic heritage of both the Hebrew and Christian Scriptures. Further, with the rise to prominence of seminal Third World theologians, the Catholic Church, always universal in theory, has in fact become international in scope. The implications are still to be realized, but they are profound.

The threat to the traditional church is not merely the political and social thrust of the Third World message. The challenge lies deeper, in the questioning of the most basic assumptions Catholics have traditionally made about the entire theological enterprise. Liberation theologians have turned everything upside down. Whereas natural law theorists hoped to disclose fundamental truths about the universe and human nature from which rules for behavior might be deduced, the new theology starts with *praxis*—right conduct—which inspires the faithful to critical reflection. *Praxis* is no longer the "goal" of theorizing, but its very ground. In the words of Gustavo Gutiérrez, "The understanding of the faith appears as the understanding not of the simple affirmation—almost memorization—of truths, but of a commitment, a particular posture toward life" (*A Theology of Liberation* [Maryknoll: Orbis Books, 1973]).

For liberation theologians, *praxis* means struggle against human oppression and solidarity with victims of unjust social structures. Victims are usually identified as the poor in the First and Third Worlds: blacks and other minority groups, and women. The category of "victim," however, is elastic. For example, Dorothy Soelle laments the victimization of affluent people by consumerism, and Matthew Lamb discusses the pollution of the earth as an oppression. It is only within the last generation that our human treatment of the earth has been widely recognized as a problem for theology. Now the earth is frequently considered to be in need of liberation, just as are women and the poor. This perspective, as we shall see, is taken up, if modified, in the writings of Thomas Berry.

*3) A rise to dominance of Anglo-American linguistic analysis.* Instead of talking about reality as apart from the observer, 20th century Anglo-American philosophy takes a step backward, talking about talk. The writings of the Vienna Circle in the 1920s asserted scientific discourse to be the norm: religious statements being empirically unverifiable were labeled meaningless. But since then, a number of noted philosophers, concentrating on the uses and contexts of statements, have challenged positivist approaches to language, showing the wide diversity of functions that statements serve. Religion has been "rescued" by studies that emphasize the ethical, ritual, command, storytelling, and aesthetic functions of religious statements.

However, linguistic analysts are likely to insist that science and religion, being distinct worlds of discourse, must not be confused. If religion, then, is impervious to scientific or philosophical attack, it is nevertheless confined to the realm of the personal and existential. The linguistic perspective provides no support for the efforts of natural law theorists to treat nature as a source of revelation in regard to human norms of behavior.

## *WHERE BERRY FITS IN*

In his "Twelve Principles for Understanding the Universe and the Role of the Human in the Universe Process," Thomas Berry asserts: "The universe, the solar system, and the planet earth in themselves and in their evolutionary emergence constitute for the human community the primary revelation of that ultimate mystery whence all things emerge into being."

Reading this assertion, we are apt to feel that we are closer to Aristotle or Thomas Aquinas than to the contemporary religious thinkers just described. Modern doubts about the inherent intelligibility of the world or about our powers to grasp it are nowhere in evidence. The grand traditional scheme is drawn for us again: nature is not only a back-drop for human activity but a source of revelation of the ultimate mystery itself. We sense the continuity of God and nature, of nature with humans, and humans in community. In such a vision, human life assumes significance not only because humans think and wonder, but also because they participate in a wider fabric of order and coherence.

Berry's assertion, the first of his "Twelve Principles," suggests a traditional Thomistic orientation in which faith, knowledge, and experience are to be reflected upon and integrated into a vast *Summa*. For Berry, as for St. Thomas, science and theology, theology and pyschology, indeed the whole world of learning is important insofar as it leads us to a saving knowledge. All of nature is a source of revelation. But whereas for St. Thomas nature is secondary to Scripture, for Berry it is in our apprehension of physical reality that we encounter in the first instance "the ultimate mystery whence things emerge into being."

But any contemporary *Summa* must be more narrative than speculative. Perhaps it is not fanciful to see in Berry's "Twelve Principles" a deliberate effort to reconstitute through narrative of the evolutionary process a new paradigm of reality and value which stands beside and implicitly comments on the metaphysical and cosmological assumptions which ground St. Thomas' work. It is notable that for Berry there is a trinity: differentiation, subjectivity, and communion, which is nature's constitutive principle and which manifests itself on every level, including that of the human individual (Principle four). But whereas for Thomas Aquinas, these are metaphysical categories, for Berry they are descriptive ones arrived at through empirical observation. It is apparent that Berry, like St. Thomas, considers things on a large scale. In his second Principle Berry tells us what the universe is, in his sixth what human beings are, and in the seventh what the earth is. In Principle eleven he indicates the sequence of historical periods, and in the twelfth he points to the historical human task that follows from the others and describes the means to achieve it.

### *SIMILAR BUT DIFFERENT*

However, while these similarities between Berry's principles and St. Thomas' principles are impressive, it is obvious that what we find in Berry's work is something different from a simple up-dating of an essentially medieval world view. It is the novelty which is apt to strike us, even overwhelm us, rather than the genuine link to an older tradition. It is a shift from the metaphysical and deductive to the historical, descriptive, and empirical.

The characteristic turn in Berry's thought is revealed in a single word: emergent. Whereas for St. Thomas, eternal laws governed a seasonal ever-renewing world based on hierarchical principles, for Berry the universe is and always has been an irreversible historical developmental process. Berry, of course, is not the first religious thinker to assimilate the theory of evolution into his works.

In fact his vision has been informed by the writings of Teilhard de Chardin, whom he greatly admires, and, in consequence, by Henri Bergson. Bergson attacked the mechanistic approach to nature and endorsed an evolutionary perspective in which life strives to free itself from the domination of matter to achieve self-consciousness. He was a strong influence on Teilhard. In both French writers one can find evidence of the influence of neo-platonic ideas, manifest especially in Teilhard's assertion that order ultimately depends on unity, and in his belief in a progressive intensification of consciousness within matter, reaching a high point first in humans, and, ultimately, at the Omega point to which all matter is being drawn.

In regard to human apprehension of the process, intuition becomes very important, as it was for the neo-platonists. Teilhard claims, and Berry reiterates, that rather than viewing nature as an object one should appreciate the subjectivity of natural objects. This preoccupation with the "within" of things, characteristic of Teilhardian thought, also has its parallels in Heidegger's work, which has been seminal for 20th century philosophy and theology.

Is Berry, then (following in the steps of Teilhard and others), offering a model of "disclosure" typical of traditional theology and found inadequate by liberation theologians? In some ways he is. According to Berry, it is by engagement with the universe process that humans discover the truth about who they are. Indeed, the human is precisely that being in whom the natural world reflects on itself. There is no suggestion of an inherent incommensurability between natural processes and our human apprehension of them. Such a dichotomy makes no sense, since reason—or more broadly, human consciousness—is integral to the emergent cosmic process.

If Berry's model is largely a disclosure model it does not follow, however, as the liberation theologians would have it, that, in this scheme, *praxis* follows theory. Once again, in Berry's work, the distinction between these traditional polarities breaks down. Even for Thomas Aquinas, reasoning was an activity; for Berry, our human participation in the universe's unfolding is manifestly an activity in which humans engage holistically. If the human is the being in which the universe becomes conscious of itself, the achievement is not strictly or only rational. For Berry, humans are genetically mandated to invent a cultural coding in which "specifically human qualities find expression " (Principle nine).

Culture in the broadest sense is determined by its sense of reality and of value. It includes manner of life, the organization and execution of all human activities; it is everything that people do.

### *THE BASIC ISSUE*

To say that humans are culturally as well as genetically coded is, of course, to recognize the important difference between the biological and intellectual realms, as well as perceiving the continuity between them. This admission, which presumably any observer would make, poses the basic issue for Berry as it did for St. Thomas. The paradox at the heart of human existence becomes apparent at the cultural level. Flowers and beasts, rocks and stars, have no choice about behaving as they must (as Aristotle would say, about behaving in such a way as to achieve their form), but humans do. While cultures generally encode behavior which insures survival—while they as a rule are the ground in which humans become "the understanding heart of the universe" (Principle six)—they sometimes cause serious injury to the planet and, ultimately, to themselves. For Berry, the rise of mechanistic science and the total subjection of the earth by industrial processes have brought about destruction on an unparalleled scale, much of it irreversible. The solution cannot be modest or piecemeal. On the contrary, what Berry calls for is "the reinvention of the human at the species level."

St. Thomas also recognized that humans can fail in understanding and not fulfill their role in the universe. Fortunately, they have a corrective at their disposal in the Divine Word as revealed in Scripture and Tradition, although these saving resources can also be misunderstood. More accurately, for St. Thomas, primary revelation comes through the Word and the teachings of the magisterium; nature is a helpful if subordinate aid. Berry, on the other hand, focuses on the message which the universe reveals to us.

Although Berry is obviously familiar with St. Thomas' work and has been influenced by Teilhard de Chardin, he does not, like them, emphasize the saving power of Christian revelation. He is not concerned "exactly" with the spiritual salvation of souls, but with the integral relationship between the human community and the planet earth, and with the integral survival of both in a mutually enhancing relationship. Berry's work is suffused with a sense of the divine; nevertheless he does not quote Scripture or church documents or speak of God in explicating his meaning.

### *DIFFERENCES BETWEEN TEILHARD AND BERRY*

Whereas Teilhard was trying to reconcile Christian belief with scientific data, Berry is trying to re-evaluate the resources available as we work to insure survival. He is not concerned with interpreting a supernatural revelation. Teilhard was optimistic about the course of cosmic history and our human role in it. He believed that the goal of evolution was complex consciousness (Omega, or Christ), and that humans were creatures eminently capable of self-reflective consciousness. Thus, he believed in the rightness of human domination over nature and felt gratified by the achievements of human culture.

Berry, on the other hand, writing in the ecological age, is dismayed at what people have wrought. Contemporary civilization must be changed before humans succeed in destroying themselves and all of nature. Christianity, embedded as it is in traditional western culture (indeed, being a leavening agent in it), must face the fact that its saving stories are presently dysfunctional: They have become disembedded from the natural processes which sustain all of life and make culture possible.

It is at this point that Berry's writings are apt to remind us of the efforts of the linguistic philosophers who study how language, in this case religious language, is used. In particular, Berry examines our archetypal stories that make our universe meaningful for us and provide the guiding models for a worthy human life. Since the 14th century, he claims, our archetypal stories have oriented us toward an excessive emphasis on redemption, on the personality of the Savior, on interior spiritual processes. While such an approach sustained us for a long time, providing a context in which life could be lived meaningfully, it can continue to do so only if this salvific role is restated within our new sense of the universe and how it functions.

All of Berry's writings may be seen as an effort to write a new story. Clearly, to be adequate, it must correct emphasis on the redemptive story to which we have become accustomed by re-embedding it in the wider communities of globe and cosmos. It must integrate a cosmology which embraces historical and scientific learning about process and change with our deepest personal and spiritual concerns.

At the same time, however, the new story will challenge the assertions of linguistic philosophers and scientists who claim that science and religion represent separated and irreconcilable worlds of discourse. Indeed, such a claim is part of the problem: there is no in-depth communion between the scientific and religious communities on "cosmic-earth-human" values. Currently both the scientific and the religious traditions are trivialized, and the human venture remains at an impasse.

Writing the new story is in many ways like writing a new *Summa*, this time as historical narrative rather than as metaphysical synthesis. In this sense Berry's project is more biblical and less hellenic in form than is St. Thomas'. The journey story of the Bible has to be retold in a larger cosmological context as the journey of the universe. Not the *Summa* of Thomas Aquinas, but rather *The City of God* of Augustine, Berry claims, could provide an appropriate model. This new narrative, besides revealing the story of the universe in all its phases, will re-enchant us with the world and with our role in it.

Part of the immense effort it will involve to tell the "new story" will require the assimilation of our more recent scientific and ecological perspectives. Another and equally important part of this effort will require reappropriating our religious past, and looking anew at our traditional insights into the human place in nature, which can help us find our way out of our present impasse. As Berry suggests, our project must be to rethink our words. We need to find in them, or forge out of them, the meaning of making ourselves into new creatures, citizens of the global community, whose deepest longings and passionate strivings will be in harmony with the needs of the larger organism of which we are the thinking part.

## *QUESTIONS FOR DISCUSSION AND REFLECTION*

1. In this paper Richards summarizes current Catholic theological thinking in three modes. In what ways is this helpful for your understanding of the contemporary literature? Do you find other major themes that she does not mention?

2. The nature of human devastation on the planet is the starting point for Berry's work. In this he shares more in common with scientists and ecologists than with theologians. Yet Richards writes that Berry's work is "suffused with a sense of the divine." In his writings in this volume, have you found this to be true?

3. The time-developmental nature of Berry's work is both reminiscent of Teilhard de Chardin and different in some crucial areas. What are these areas?

4. Richards writes: "Berry's project is more biblical and less hellenic in form than is St. Thomas' [*Summa*]. The journey story of the Bible has to be retold in a larger cosmological context as the journey of the universe." In what ways does this surprise you? Does it fit in with your reading of Berry's work?

5. Richards also mentions that Thomas Berry claims a link with Augustine's *The City of God*. How would you explain this link?

# 10

# *OUR FUTURE ON EARTH:*

## *Where Do We Go From Here?*

Thomas Berry

We hope that the essays in this book have helped clarify some of the urgent issues that are of concern not only for humans but for every living being on the planet. This clarification is a first step toward becoming aware of the peril facing the earth and all its living creatures. We begin to recognize our responsibility for the fate of the earth, both by ceasing our industrial assault upon it and by initiating programs for the renewal of those life systems that have been severely damaged. Thus far, however, our biblical, our social, our ethical, and our theological traditions have provided only minimal resistance to the degradation of the earth. That there are powerful resources within these traditions to aid in this renewal process we can be certain. To involve these resources fully is the immediate task that is before us.

In a recent conference on species diversity sponsored by the National Academy of Science and the Smithsonian Institution (in Washington, D.C.), there was wide agreement that only in the great geological and climatic upheavals of the past could we find any parallel with the present rate of species extinction (caused by human activity). A speaker at the conference, Norman Myers, considered that the impending situation would likely produce "the greatest single setback to life's abundance and diversity since the first flickerings of life almost four billion years ago." And speaker Paul Ehrlich suggested that we could be bringing upon ourselves consequences "depressingly similar to those expected from a nuclear winter, including famine and epidemic disease."

Obviously we are not simply in a period of historical change or cultural modification. We are in a situation beyond anything ever experienced before in the course of human or earth history. Though the situation may not have been "caused by" our biblical or theological or ethical or social programs, these have not been able to prevent or critique the situation in any effective manner, and they are not presently offering adequate guidance.

Beginnings are being made, however. A book like this is certainly one such beginning. We might hope that more and more effective efforts will continue to be made to reread the story that the universe tells of itself. Genesis began with an ancient perception of the universe. We, too, must begin with our perception of the universe. Though the traditional world assumes that the earth and its living forms are abiding and ever-renewing realities, we experience the earth as emergent and irreversible process. We are vastly powerful, so powerful indeed that we are definitively closing down many of the basic life systems of the planet through our technological skills.

### *OUR BEST RESOURCE IS EARTH*

Our primary resource in this situation, it seems to me, is not our biblical revelation but the revelation granted us through the universe, the planet earth, the living world about us, and the course of our own human development. So too with our moral teachings and our social bonding; these are considerably deepened and extended by recognition that the universe is a psychic-spiritual as well as a physical-material reality from its beginning. It is integral with itself throughout the full extent of space and the total sequence of its temporal transformations. We are, by definition, that being in whom the universe reflects on and celebrates itself in conscious self-awareness. We are the universe in its self-awareness phase. In and through this universe-identity we have our identity with that numinous mystery whence all things emerge into being. Our alienation from the divine, our alienation from the natural world, and our alienation from ourselves and from each other, all these are different aspects of a single reality.

What we do to the universe in its earth expression we do to ourselves in the larger mode of our being. The future depends on our ability to establish a mutually enhancing relationship with the natural world about us and with all its component members. In achieving this we need to recognize that the larger context of existence and of life is the primary reality to be preserved, just as the well-being of the larger organism is prior to the well-being of any of the components parts of the organism. The arm fits into the body, the body does not fit into the arm. In the proper course of affairs these are mutually enhancing.

So too there is need to remember that the human is a component member of the earth and its life systems. The earth and its well-being are our basic referent as regards reality and value. The earth can exist without us. We cannot exist without the earth. In the design of nature there exists a mutually enhancing relationship. Our difficulties arise from our efforts to make the earth subservient to our phenomenal ego rather than to discover our true grandeur by fulfilling our role within this larger scheme of things.

There is, finally, one subject of considerable import that must not be neglected, the subject of patriarchy. While this is a gender issue, it is also a pervasive and surely a controlling aspect of the entire earth-human process. Whatever has happened in the human process and especially in the western cultural process has been profoundly affected by masculine dominion over the basic institutions that have governed our history.

The four patriarchal institutions that have governed western history could be listed as the political empires, the institutional church, the nation-state, and the modern corporation. A history of western civilization could be written in these terms. That much has been achieved within these institutions is clear. But the magnitude of the oppression and destructiveness of these institutions has only recently been recognized. As these establishments endure, the consequences of their exploitation may soon be something akin to nuclear winter. Their patriarchal plundering processes are devastating the natural systems of the planet.

If there is to be any acceptable future for the variety of living forms that constitute in great part the splendor of the earth, or if there is to be any acceptable human future, the grandeur of this planet must continue to flourish. This can only come about by a transformation of patriarchal dominion to a more nurturing attitude, both toward the natural world and all its living creatures, and of humans toward each other.

# *TWELVE PRINCIPLES:*

## *For Understanding the Universe and the Role of the Human in the Universe Process*

Thomas Berry

1. The universe, the solar system, and the planet earth in themselves and in their evolutionary emergence constitute for the human community the primary revelation of that ultimate mystery whence all things emerge into being.

2. The universe is a unity, an interacting and genetically-related community of beings bound together in an inseparable relationship in space and time. The unity of the planet earth is especially clear; each being of the planet is profoundly implicated in the existence and functioning of every other being of the planet.

3. From its beginning the universe is a psychic as well as a physical reality.

4. The three basic laws of the universe at all levels of reality are differentiation, subjectivity, and communion. These laws identify the reality, the values, and the directions in which the universe is proceeding.

5. The universe has a violent as well as a harmonious aspect, but it is consistently creative in the larger arc of its development.

6. The human is that being in whom the universe activates, reflects upon, and celebrates itself in conscious self-awareness.

7. The earth, within the solar system, is a self-emergent, self-propagating, self-nourishing, self-educating, self-governing, self-healing, self-fulfilling community. All particular life systems in their being, their sexuality, their nourishment, their education, their governing, their healing, their fulfillment, must integrate their functioning within this larger complex of mutually dependent earth systems.

8. The genetic coding process is the process through which the world of the living articulates itself in its being and its activities. The great wonder is the creative interaction of the multiple codings among themselves.

9. At the human level, genetic coding mandates a further trans-genetic cultural coding by which specifically human qualities find expression. Cultural coding is carried on by educational processes.

10. The emergent process of the universe is irreversible and non-repeatable in the existing world order. The movement from non-life to life on the planet earth is a one-time event. So too, the movement from life to the human form of consciousness. So also the transition from the earlier to the later forms of human culture.

11. The historical sequence of cultural periods can be identified as the tribal-shamanic period, the neolithic village period, the classical civilizational period, the scientific-technological period, and the emerging ecological period.

12. The main human task of the immediate future is to assist in activating the inter-communion of all the living and non-living components of the earth community in what can be considered the emerging ecological period of earth development.

# *SELECTED READINGS*

*The Riverdale Papers*, Vols. 1-10 and continuing. Thomas Berry, Riverdale Center, 5801 Palisade Ave., Bronx, NY 10471.

## *SCIENCE*

*Gaia: An Atlas of Planet Management.* Ed. Norman Myers. New York: Doubleday, 1984. A comprehensive study of our living planet at a critical point, as one species threatens to disrupt and exhaust its life support systems. Interdisciplinary. To be revised and updated periodically.

*Timescale: An Atlas of the Fourth Dimension.* Nigel Calder. New York: Viking Press, 1983. First book to combine complete graphic and verbal depictions of our cosmos in time, from the Big Bang to the space shuttle. Fun to read.

*Earth and Life Through Time.* Steven Stanley. New York: Freeman and Co., 1986. A reference book, somewhat like Calder's, relating geology, climate, biology, and history in a journey through time on the earth.

*The Anthropic Cosmological Principle.* John D. Barrow and Frank J. Tipler. Oxford: Clarendon Press, 1986. The history of philosophic thought concerning the question of design and humans' place in the universe is investigated. The modern collection of ideas known as the Anthropic Cosmological Principle emerges historically as the latest manifestation of such ideas. It is highly technical but readable; very expensive.

*The Lives of a Cell.* New York: Bantam Books, 1973. *The Medusa and the Snail: More Notes of a Biology Watcher.* New York: Penguin, 1979. *Late Night Thoughts on Listening to Mahler's Ninth Symphony.* New York: Penguin, 1983. All by Lewis Thomas. Thomas' collections of essays from the New England Journal of Medicine are beautiful words on the astounding universe.

*The Immense Journey.* Loren Eiseley. New York: Time, 1962. The classic by "the great humanist among scientists."

*The Soul of the Night: An Astronomical Pilgrimage.* Chet Raymo. Englewood Cliffs, N.J.: Prentice-Hall Inc.,1985. "A personal pilgrimage into the darkness and the silence of the night sky in quest of human meaning."

*The Science Question in Feminism.* Sandra Harding. Ithaca, N.Y.: Cornell University Press, 1986. This is a far-ranging critique of the fundamental assumptions of science in the west.

## *PHILOSOPHY AND PHILOSOPHY OF ETHICS*

*The Embers and the Stars: A Philosophical Inquiry into the Moral Sense of Nature.* Erazim Kohak. Chicago: University of Chicago Press, 1984. "In philosophical terms, Kohak proposes a personalist inversion that treats the personal order of the cosmos as ontologically and ethically primary." A beautiful and meditative book.

*The Imperative of Responsibility: In Search of an Ethics for the Technological Age.* Hans Jonas. Chicago: University of Chicago Press, 1984. This work is a comprehensive beginning for a new ethics based on the present crisis in earth/human relationships. Technical, but very fine.

*The Modern Crisis.* Murray Bookchin. Montreal: Black Rose Books, 1987. This is a fine collection of essays by someone who has been most important for the bio-regional concept.

## *GENERAL LITERATURE*

*Not Wanted on the Voyage.* Timothy Findley. New York: Penguin, 1984. One major theme of this novel is a questioning of the dark side of our religious tradition and our western alienation from nature. It is a magnificent synthesis of feminist and ecological issues.

*Pilgrim at Tinker Creek.* New York: Bantam Books, 1974. *Holy the Firm.* San Francisco: Harper & Row, 1977. *Teaching a Stone to Talk.* New York: Harper & Row, 1983. All by Annie Dillard. Dillard combines meditations on the cosmos with a profound theological vision.

## *HISTORY*

*The Western Spirit Against the Wilderness.* Frederick Turner. New York: Viking Press, 1980. This is a controversial, eloquent exploration of the role of Christianity in the devastation of peoples, notably North American native populations. It is helpful background for the myth of progress and western attitudes that still exist in such forms as "manifest destiny."

*The Death of Nature: Women, Ecology and the Scientific Revolution.* Carolyn Merchant. New York: Harper & Row, 1980. This careful study of the rise of science and its results in the last few centuries is important from a feminist and ecological perspective on our present world.

*Soil and Civilization.* Edward Hyams. New York: Harper & Row, 1952. Reprinted 1976. A history of the vicissitudes of the earth, once humans began cultivation. Very easy and absorbing reading.

*Nature's Economy: the Roots of Ecology.* Donald Worster. New York: Doubleday, 1977. A history of ecology from the 18th century on.

*Dwellers in the Land: The Bioregional Vision.* Kirkpatrick Sale. San Francisco: Sierra, 1985. A fine introduction to the bio-regional concept, with extensive suggestions for reading in history and present works.

## *RELIGION AND COSMOLOGY*

*Green Paradise Lost.* Elizabeth Dodson Gray. Santa Monica: Roundtable, 1979. Feminist and ecological concerns lead Gray to "remything genesis." A very readable introduction to some of the major questions raised about Christianity in relation to our alienation from nature.

*The Universe Is a Green Dragon.* Brian Swimme. Santa Fe: Bear and Co., 1983. This popular book by a physicist is a fascinating exploration of the implications of present cosmological thinking.

*The Cosmic Adventure: Religion, Science and the Quest for Purpose.* John Haught. Mahwah, N.J.: Paulist Press, 1984. Many concerned with religion and science jump too quickly into drawing from science much more than it intends with quantum physics, etc. This is a careful account of Whitehead's and Polanyi's thought. Haught only suggests an openness to religious insight and the new cosmology. He also tries to deal with the question of evil in a new, thoughtful framework.

*Ethics in a Theocentric Perspective*. Vols. 1 and 2. James Gustafson. Chicago: University of Chicago Press, 1982 and 1984. These volumes that Gustafson considers the culmination of his ethics are written with the new cosmological framework as basis. The Calvinist emphasis is strong in this first systematic theological ethics in the new vein.

*God in Creation: A New Theology of Creation and the Spirit of God.* Jürgen Moltmann. San Francisco: Harper & Row, 1985. Moltmann, who has influenced so many present themes in theology, notably liberation theology, writes with the perspective of the present crisis of life on earth. He plumbs some of the theological riches of the tradition, notably the idea of sabbath, jubilee, and trinitarian thinking.

*Theology for a Nuclear Age*. Gordon Kaufman. Philadelphia: Westminster, 1985. This readable, short volume raises the problem of our traditional understanding of providence and humanity in a situation that our heritage was not meant to address. It situates the problem very clearly.

*Imaging God: Dominion as Stewardship*. Douglas Hall. Grand Rapids: Eerdmans, 1986. Hall, like Berry, accepts some Christian responsibility for the present ecological crisis, and delineates the themes in our history that contribute to it. He continues with a reinterpretation of the "imago Dei." Clearly written and engaging. Like Moltmann and Berry, Hall sees the present crisis as the most serious challenge to theology and Christian life that must be faced.

*Universe: God, Science, and the Human Person*. Adam Ford. Mystic, Conn.: Twenty-Third Publications, 1987. Ford describes how a scientific approach to the universe, as seen in astronomy and sub-atomic physics, has immense riches to offer to religious faith; faith, in turn, illuminates the world described by science.